BRITISH ARTILLERY
ON LAND AND SEA
1790-1820

BRITISH ARTILLERY ON LAND AND SEA 1790-1820

Robert Wilkinson-Latham

DAVID & CHARLES : NEWTON ABBOT

To Christine

0 7153 5448 5

© Robert Wilkinson-Latham 1973

Set in 11/13pt Plantin
by Avontype (Bristol) Limited
and printed in Great Britain
by William Clowes & Sons Limited Beccles
for David & Charles (Holdings) Limited
South Devon House Newton Abbot Devon

CONTENTS

INTRODUCTION

The year 1716 was one of the most important in the history of British artillery. On 26 May two permanent artillery companies were formed by royal warrant and stationed at Woolwich while later in the year the Royal Brass Foundry was built there.

Before 1716, the artillery had to be formed into trains in time of war and had to assemble the equipment, draught animals and drivers prior to marching. The nucleus of the trains was the force of gunners and their assistants who were stationed at the Tower of London and at various castles and garrisons elsewhere in Britain. Henry VIII was first responsible for establishing a force of artillery when he appointed Humphrey Walker as Master Gunner at the Tower of London together with twelve assistants, and although William III had established a regiment of artillery in 1697 (disbanded the following year), the artillery at the beginning of the eighteenth century was still basically formed along the lines laid down by Henry VIII. It was no longer adequate. During the 1715 uprising in Scotland, for instance, the train had not even completed its equipping before the rebellion was over. This weakness in an army taking the field only emphasised the need to establish a permanent artillery force, which was done in 1716.

In 1722, the two companies formed in 1716 were augmented by grouping them with the artillery train formed in Gibraltar in 1704 and the train formed in Minorca in 1709. Further companies were added later, including (in 1744)

the 8th company which consisted of 'gentlemen cadets' who underwent training prior to receiving their commissions from the Master General of the Ordnance. The brigade system of grouping artillery together was first established in 1756, when at the practice camp at Byfleet brigades were formed consisting of 4–6 guns of similar calibre. The new formation included three 24 pdr brigades, three 12 pdr brigades, four 6 pdr brigades and one brigade of 3 pdrs together with two brigades of Royal Howitzers. By 1757 there were sufficient companies to form two battalions to which a third was added in 1759 and a fourth in 1771. There were also eight invalid companies, two being attached to each battalion. These companies were formed from old and pensioned soldiers. In 1794 a fifth battalion was formed, followed by a sixth in 1799.

In 1801 a seventh battalion was added, which, until that date, had been the Royal Irish Artillery. Formed in 1757, the Royal Irish Artillery was originally a separate force which in 1760 came under the command of its own Master General of Ordnance, John, Marquis of Kildare. The Irish Artillery had its own Masters General until it was merged with the Royal Artillery in 1801.

The Duke of Richmond as Master General of the Ordnance had issued instructions in 1788 for the formation of a field battery that was able to keep up with cavalry, and in 1793 the Royal Horse Artillery was formed. At first this new arm of the service consisted of four troops, but they were quickly augmented by the addition of two more. By 1805 the number had risen to twelve, and in 1813 was increased to fourteen with the addition of 1st and 2nd Rocket Troops. After 1815 and the coming of peace, the number of troops was finally reduced to seven.

The establishment of men, horses and material of a typical Horse Artillery troop (G Troop) in 1815 was as follows.

Horses and material
Five 9 pdr guns and one 5½in howitzer, with eight horses to each
Nine ammunition waggons with six horses each
One spare wheel waggon with six horses
Forge, curricle cart and baggage waggon with four horses each
Horses for officers, NCOs and men, and baggage animals, totalling 106
Total complement of horses: 226
Personnel
First captain, second captain, 3 lieutenants, 1 surgeon, 2 staff sergeants, 3 sergeants, 3 corporals, 6 bombardiers, 1 farrier, 3 shoeing smiths, 2 collar makers, 1 wheeler, 2 trumpeters (one was a driver), 80 gunners, 84 drivers
Total complement of personnel: 193

Prior to 1716, guns were cast by private concerns and supplied to the Ordnance under contract. The majority of the iron guns at this time were supplied by such concerns as Fullers in Sussex, but all brass ordnance came from the Moorfields foundry in the City of London. The Moorfields foundry had been started in 1684 by Maximilian Western, who continued in business until 1704 when the foundry was leased to Mathew Bagley. At this time all unserviceable and captured guns were broken up and the metal was used for casting new ordnance, which was then proved at the butts at Woolwich before being accepted into service. The transport to Woolwich and the proving were at the contractors' expense, and there were always complaints from them because, if their guns failed proof, they received no payment, as the metal was the property of the Board of Ordnance anyway. Plans were prepared in March 1716 for the construction of a Royal Brass Foundry at Woolwich so that the Board of Ordnance would not have to rely solely on outside contractors for their supplies. The disastrous explosion which took place at the Moorfields foundry in May 1716 speeded up the decision of the government to build the foundry at Woolwich.

Even though the new foundry supplied a large amount of ordnance, outside contracts were still placed with Fullers and later with the Carron Company which, although it became more famous for its 'carronades', cast all types of ordnance.

Andrew Schalch, a 24-year-old German who had been employed as a founder at the *Fonte Nationale* at Douai in France, was appointed to organise the new foundry and became the first Master Founder. By January 1717 Schalch had started casting and soon embarked on a programme of casting 24 pdrs of which there was an acute shortage at the time. Schalch worked at the foundry until 1770, having in his time introduced many improvements in the mode of manufacture. In January 1770, now 78 years old, he was replaced by Jan and Peter Verbruggen who had been brought over from Holland.

The Verbruggens had been responsible for the change from guns cast with a bore to those cast in the solid and bored afterwards. On their arrival at Woolwich they set to work enlarging the foundry and improving a number of the methods of manufacture. A quantity of machinery was brought over from Holland and installed at Woolwich on their instructions. Jan Verbruggen died in 1782 and Peter in 1786, after which date no Master Founder was appointed at Woolwich. A number of other founders continued their work after their deaths, namely John and Henry King who worked from 1788 to 1811 and Francis Kinman who worked from 1794 to 1817, and their names frequently appear on the barrels of ordnance during the period 1790–1820.

From about 1700, pieces of ordnance had become known by the weight of the shot they took rather than the strange method of denoting each size by the name of a monster or bird of prey. Howitzers and mortars were known, not by the weight of the shot but by the diameter of the bore in inches. Typical examples of the naming of ordnance are Falcon, Demi-Cannon-Drake, Culverine, Dragon, and Murderer, a name given to a small mortar.

Once the Royal Brass Foundry had been established and Andrew Schalch had become Master Founder, a number of new types of piece came into existence. In 1719 the first 9 pdr brass gun had been cast and in the same year the first 8 inch brass howitzer had been manufactured. In 1728 a $4\frac{2}{5}$ inch howitzer had been made, followed by a 10 inch one cast in 1729. Carriages had become standard in design and consisted of a double trail connected by an axle-tree. Semicircular cut-outs in the top of each 'cheek', lined with metal, were positioned to accept the trunnions of the barrel. Once the barrel was in position it was held there by capsquares which closed over the trunnions and were held by wedges to the iron-work of the 'cheeks'. In the larger calibre guns there were additional cut-outs in which the barrel was placed when being transported, these being positioned slightly behind the main capsquares at the point of balance when the trail was attached to the limber.

The method of manufacture also altered, and guns were now cast solid and then bored afterwards rather than being cast with the bore which was more often than not in-accurately placed. Elevation of the barrels was achieved by means of a quoin or wedge which was pushed under the breech end of the barrel resting on the stool of the carriage. The more the wedge was pushed in the more horizontal the barrel became and vice versa. Later, this type of quoin was replaced by one which had a vertical screw thread between the two halves worked by a handle and which gave more accurate elevation and depression. Although this method was still used on garrison carriages in 1790 and much later it was replaced by an elevating screw on field carriages.

Artillery before 1793 was not highly mobile, owing to the fact that the gunners marched with the train of artillery to the place of action. Light artillery or the mobile use of artillery had been experimented with as early as the battle of Blenheim (1704) when the Duke of Marlborough had ordered Colonel Blood to march a battery from its position across a river to engage the enemy, and again at the battle of Fontenoy when use was made of light guns mounted on carriages pulled by a horse, known as gallopers or galloper guns. $1\frac{1}{2}$ pdr guns mounted on split trail carriages had been used briefly in 1747 but discontinued the following year. Light artillery had been used on the Continent many years before its introduction into England. The king of Prussia had a light form of artillery pulled by six horses in 1774, and this set the pattern for other countries to follow. The need for light artillery in the British and French armies had been filled by having small-calibre guns, 6 pdrs for the British and 4 pdrs for the French, as battalion or infantry guns. In the British service, a number of 6 pdrs were allotted to each battalion and manned by the Royal Artillery. This system led to a decline in the morale of the gunners, many of whom spent years attached to an infantry battalion. In 1798, and again in 1801, decisions were taken that 1 officer and 18 men of each cavalry regiment and 1 officer and 34 men of each infantry regiment were to be trained to serve the gallopers and to release the gunners for duty on the heavier pieces. However, by 1799 the use of battalion guns had virtually died out in the British service.

Perhaps the greatest development in eighteenth-century artillery, after the block trail designed by Congreve in 1792 (see page 51, below), was the carronade. The Carron Company had started manufacture of barrels and shot in 1759, and a joint design effort between Mr Gascoigne, the works manager, and General Robert Melville resulted in the short-barrelled carronade (see page 16, below), introduced into the British service in 1776.

A certain amount of standardisation had taken place in British Artillery by 1790, though there is no doubt that the French were more highly organised. In the British service the number of varieties in each calibre had been greatly reduced and definite rules had been laid down for the construction of barrels and carriages together with their proof.

The Napoleonic wars saw a great many advances in the design and use of artillery and projectiles, both on land and at sea. Some ideas when tried under battle conditions proved to be unsatisfactory—for example, the total arming of ships with carronades—but overall the picture is one of quick development in the history of artillery.

1 ORDNANCE IN USE ON LAND

There were many types of artillery in use on land during the period 1790–1820. Many more developments took place such as the formation of the Royal Horse Artillery in 1793, and the introduction of Congreve's block trail and of better designed carriages and new shells. The Navy remained fairly stagnant in armament design during this period, as will be seen in Chapter 2; but the Royal Artillery progressed with its new developments in both design and tactics.

A large variety of pieces were in service at the beginning of the Napoleonic wars and each type and size had a specific purpose, whether field, siege, light, or garrison and coastal artillery or whether the pieces were guns, mortars, howitzers or carronades.

The land service had many more types of ordnance than the Royal Navy, but during the rigours of the war many of the larger brass guns were found to be wanting and were declared obsolete. Much weeding out was done amongst the remaining calibres and a number of lengths were made obsolete. In the iron guns, only the smaller calibres of 4 and 3 pdrs were made obsolete, the rest continuing in service with hardly any variation at all.

During the Pensinsula War, the 10 inch and 8 inch brass howitzers failed and were withdrawn. It is interesting to note that these were first cast in 1727 and 1719 respectively and had

1 The general construction of brass field howitzers

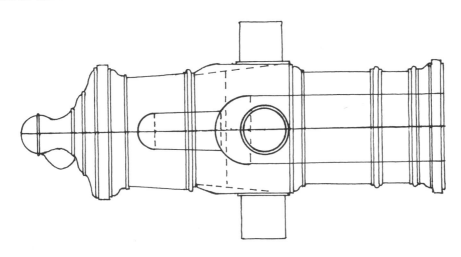

2　Carronade mounted on a wooden rear chock garrison carriage

undergone hardly any change in design. The main howitzer armament consisted of the heavy and light $5\frac{1}{2}$ inch and $4\frac{2}{5}$ inch which continued in service throughout this period. The deficiency in the larger calibres of howitzer was made good by the introduction of a 10 inch and an 8 inch made in iron. These were in use towards 1820.

The land service use of carronades was mainly in fortresses for flanking fire and in confined areas in garrisons. Mortars still continued in service in a large variety. There were six brass land service and four iron land service including a stone mortar which was capable of only short ranges. (This fired stones in place of cast shot.)

Except for the block trail introduced by Congreve in 1792 and used on the 3, 6 and 9 pdr guns, the carriages for the rest of the ordnance remained the same throughout the period. There were a number of refinements added but the basic design was unaltered.

A novel type of garrison carriage was designed by Lieutenant G. F. Koehler which was used in the siege of Gibraltar in 1781. It was found that the guns could not be depressed sufficiently to be able to engage the close-in enemy ships and that the shot from the garrison-carriage-mounted ordnance was going over the enemy. Koehler's design enabled the guns to fire downwards on the Spanish fleet but did necessitate the use of wads to keep both the charge and projectile in the barrel.

An advance had been made in the land service with elevating gear for ordnance, which in the sea service was not used (except for carronades),

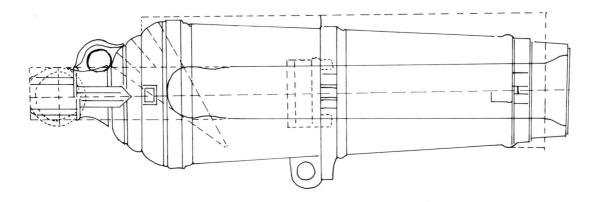

3 *The general construction of carronades*

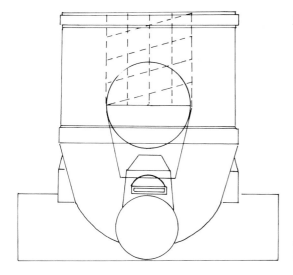

4 *The general construction of 13in, 10in and 8in iron land service mortars*

the older method of the quoin being kept to. On field carriages an elevating screw was fitted to the carriage and to the neck of the cascable. The larger garrison guns and mortars still used quoins, except in calibres of guns that proved too heavy for this type of elevating gear. The use of the elevating screw was confined to the field carriage. A further development of this, which came about in the mid 1820s, was for the elevating screw to be detached from the cascable and just to rest under it.

At sea, the carriages were kept their natural wood colour or occasionally painted, but on land all carriages were painted. The colour was a greenish grey and was known as the common colour. The ironwork and the barrels of iron guns were painted with a mixture of black with a little red paint added. The bores of the guns were lacquered to preserve them. The formula for the mixture was: 36 oz Cumberland black-lead, 1 gallon linseed oil, 10 oz red-lead and 1 oz lamp black.

Throughout the period 1790–1820, only the Royal Horse Artillery (from 1792, when they were introduced), rode horses with the guns, the rest of the Artillery still marching with the guns.

The mobile part of the artillery consisted in the main of the Royal Horse Artillery and the battalion guns. However, there was another

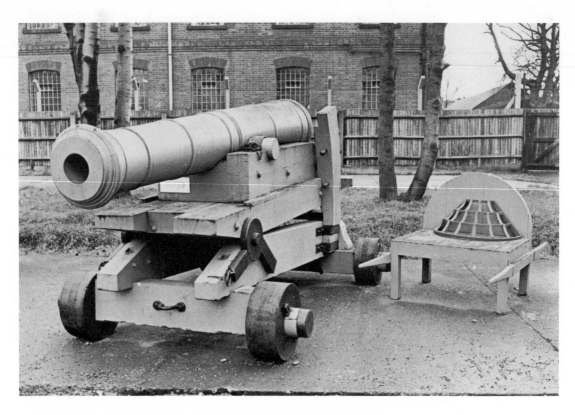

5 24pdr gun mounted on a 'Gibraltar' depression carriage, with short carrier

form of mobile artillery, Sadler's 'War Chariot'. In 1798 Mr Sadler of Pimlico, who also had his own body of volunteer sharpshooters, had his war chariot depicted in a print by Rowlandson. The carriage was designed to be extremely mobile, to keep up with cavalry and to be able to fire on the move in all directions, forwards, backwards and to each side. The range of the 3 pdr guns was said to be two furlongs. The gunners, two in number, with a third as driver, could handle the two guns 'with alacrity and in safety'.

Another form of mobile artillery was Rocket Troop. This had been formed by order of the Prince Regent on 1 January 1814. The men were equipped like the Royal Horse Artillery except that in the troop they had ammunition horses and rocket cars and not guns and limbers. There were two holsters carried by each man, with two 12 pdr rockets in each holster. The sticks for these rockets, which were seven feet long, were carried in a bundle on the off side of the horse, the thicker ends in a bucket suspended from the pommel and the sticks laid across the gunner's thigh and under his arm. The rockets in their holsters were covered with a shabraque which gave the man the appearance of a lancer. This was obviously intended to fool the enemy and surprise him when the rockets came into action. The gunner also carried a small spear-head by which he could transform one of the rocket sticks into a lance.

The types of ordnance used in the Royal Artillery during the period 1790–1820 are listed in Appendix 2.

6 *The painting* Guns to the Front *by W. B. Wollen shows the Royal Horse Artillery coming into action in the Peninsula War. The sergeant on the right of the picture is shouting orders at the lead driver of the gun. The horse harness, the uniforms and equipment are well illustrated, as is the gun and limber*

7 *Sadler's Flying Artillery, from a print by Rowlandson, 1798. This shows one of the attempts to form a mobile light artillery*

2 ORDNANCE IN USE AT SEA

The ordnance used in the Royal Navy were all of iron except for some mortars aboard the bomb ketches. They consisted of nine calibres —42, 32, 24, 18, 12, 9, 6, 4 and 3 pdrs—plus $\frac{1}{2}$ pdr swivel guns. The definition of the guns was calculated from the weight of the balls they fired. In the land service, the different calibres were usually of a fixed length, while in the Navy the lengths differed according to the place on the ship where the piece was designed to go. This was so that the guns would clear rigging and fit into corners of decks.

8 Carronade barrel showing block for fitting sight and vent field with position for cannon lock

Because of the variety of lengths in the same calibre, few guns at sea were properly proportioned to the size of shot they fired. If, for example, a 24 pdr was $9\frac{1}{2}$ feet long and twenty-one diameters of its shot in length, a 6 pdr should be in the same proportion. However, most of the 6 pdrs were thirty-one times the diameter of their shot in length, that is to say, nine feet.

The carriages on which ordnance was mounted at sea varied hardly at all during the period 1790–1820.

Carronades had been introduced in 1779. They were first made at Carron near Falkirk in Scotland. They were designed to supplement

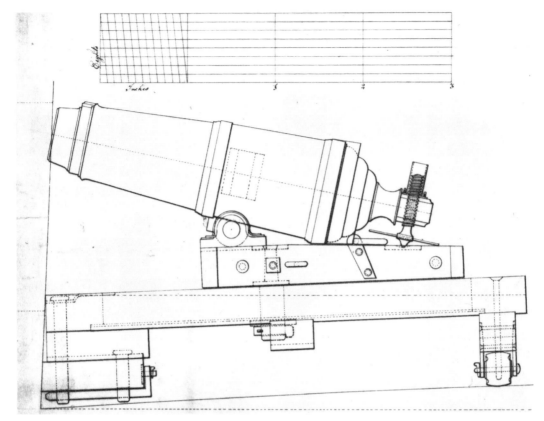

9 *18pdr joint carronade as mounted on board ship*

the larger armament of ships and to be used in confined spaces both on land and at sea. They had no trunnions and were secured to their slide carriages by a loop cast underneath the barrel. They fired a large shot and were hollowed out at the muzzle for ease of loading and so that any rigging could not be damaged by flash. They had a smaller detachment than any other piece and although not possessing great range were considered very useful for short-range work. These guns were made in seven calibres—68, 42, 32, 24, 18, 12 and 6 pdr—although the 6 pdr does not seem to have been introduced until later than 1779 and was made obsolete in about 1812.

During the Napoleonic war a number of small ships were armed solely with carronades, but the war of 1812 showed that this was a mistake. In the engagement between the Lake Erie squadron and the Americans in 1813, Sir James Yeo found it impossible to get close enough to engage the enemy with his carronades. He stated in his despatch:

We remained in this mortifying situation for five hours with only six guns in the fleet that would reach the enemy. Not a carronade was fired.

Captain Barclay stated in his letter of 12 September:

The other brig of the enemy, apparently destined to engage the *Queen Charlotte*, supported in like manner by two schooners, kept so far to windward, as to render the *Queen Charlotte's* 24 pdr carronades useless, whilst she and the *Lady Prevost* were exposed to a heavy and destructive fire from the *Caledonian* and two other schooners, armed with long and heavy guns.

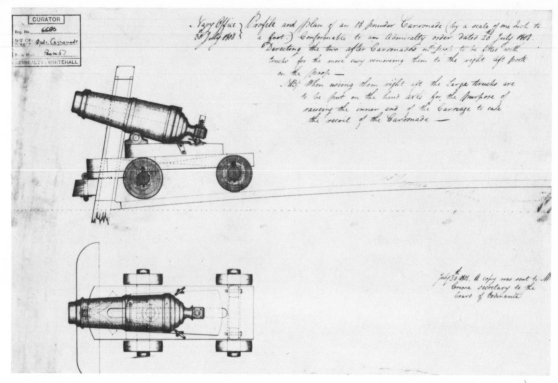

10 *Carronade mounted on a carriage with trucks, 1809*

Carronades were normally mounted on traversing slide carriages, but certain numbers were mounted on carriages and trucks. An Admiralty order of 20 July 1808 stated that the two after carronades on the poop were to be fitted with trucks for the more easy removing to the right aft port on the poop. A note was appended to the plan to the effect that if the carronades were used on the right aft, the large trucks were to be put on the aft and vice versa. Another carronade carriage was that mounted on board a river barge at Woolwich, 'agreeably to General Bentham's directions', in 1796. The carronade was cast with trunnions and the carriage was more like a garrison carriage which traversed on a lower carriage.

In 1756 it was proposed to place howitzers on board ships, but this was objected to, unless artillerymen could be sent on board to service and fire them. This had been suggested after an experiment at Gibraltar in 1736 when a shell from a 10 inch howitzer pierced a target made of fir three feet thick and went five feet into the bank behind at a range of 150 yds. An incident happened during the American war when the American ship *Deliverance* was sent up-river to bombard Philadelphia. She was forced to strike her colours after being hit by a shell from a howitzer. It was also noted that the *Deliverance* had a number of 3 pdr howitzers mounted in her tops for firing grape.

The idea of mounting howitzers was dropped but in 1760 several experiments were tried on Acton Common, firing shells from 12 and 24 pdr guns to be applied to sea service. Unfortunately it was found that the guns tended to burst and were therefore too dangerous for use on board ships. Oblong shells were also tried in a 42 pdr but with no better success.

The armament of the Royal Navy did not change considerably during this period and the methods of elevating and firing remained as they had been in 1790. A certain amount of standardisation had taken place before 1790 and during the period a number of new variations of pieces were added. The Navy relied on its heavily armed ships for naval engagements and on the bomb ketches for in-shore work.

The gunnery on board a ship of the line in 1805 is admirably shown in the Dennis Dighton painting of the death of Nelson. The painting shows a number of short 12 pdr quarter-deck guns in action. A variety of drill is taking place. The gun in the left foreground is being rammed and the lock primed while the gun next to it is being fired under the direction of a midshipman with a lieutenant looking on. On the right, men are traversing the gun with handspikes, while at the bow of the ship, a lieutenant is supervising the gunnery. A powder man who has just been shot can be seen carrying a charge in the middle foreground, and beside him there is a match tub with the match hanging over the edge. The central figure in the foreground is Midshipman Pollard, returning the enemy's fire with a musket after the Admiral has been mortally wounded. Pollard claimed to be the man who shot the man who shot Nelson. The painting also shows the uniforms worn at the time.

The types of ordnance in use in the Royal Navy in about 1790 and in about 1820 are listed in Appendix 3.

11 32pdr carronade fitted with trunnions according to the design of General Bentham, 1796

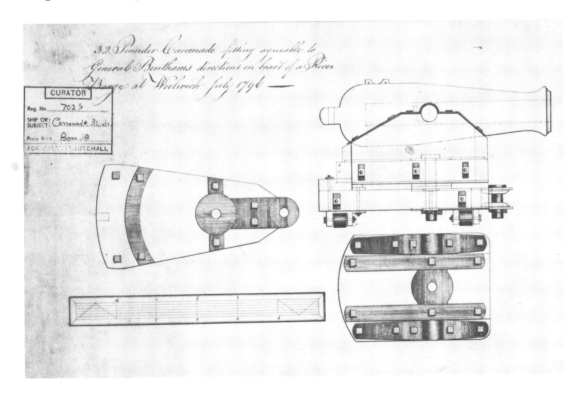

12 The death of Nelson, *by D. Dighton.*
The various guns can be seen in action and in the
course of loading

3 POWDER, SHOT AND SHELL

Gunpowder

Gunpowder is composed of saltpetre, sulphur and charcoal, its strength and quality depending on the saltpetre being well refined and the whole composition being thoroughly mixed so that in any sample, however small, there is a proper proportion of the three different materials.

The Honourable East India Company was, by its charter, obliged to provide the Government with a quantity of petre every year at a fixed price. The raw material was delivered from the government store to the powder-makers, and from every hundredweight of petre they were able to extract between six and eight pounds of saltpetre. If the petre was not properly refined before being mixed with the other ingredients, the gunpowder tended to become moist and therefore of little use. The petre was usually imported in bags containing 154 lb each and was referred to by the powder-makers as 'grough petre', being further distinguished by the terms 'private petre' and 'Company petre' depending on who was the importer. Private petre was considered the purest, containing only between 3 and 6 per cent impurities, while the Company petre contained anything from 10 to 30 per cent. By the terms of its charter, the Honourable East India Company supplied petre at £38 10s od per ton in time of peace and at £45 per ton in time of war.

Gunpowder was naturally of paramount importance to any army, and large quantities of it were used. At the siege of Ciudad Rodrigo in January 1812, 74,978 lb of gunpowder were consumed in 30½ hours, while at the storming of Badajoz in March 1812, 228,830 lb of powder were used in 104 hours. Of the gunpowder manufactured in each year, private contractors accounted for between 8,000 and 10,000 barrels in time of peace and between 10,000 and 14,000 in time of war. A barrel contained 100 lb of powder.

The amount of powder received into the Royal Magazines between 1776 and 1782 amounted to 244,349 barrels, being an average of 3,490,700 lb a year. In 1783 there was in store in the various magazines in Great Britain, Guernsey, Jersey and the Isle of Man a total of 80,000 barrels of gunpowder.

It is of interest to note that, during the 1790s, at least 30,000 barrels of gunpowder were stored in the basement of the White Tower, directly under the public records.

The usual proportions for 100 lb of Ordnance powder were 80 lb of refined salt-petre, 20 lb of sulphur, and 10 lb of charcoal; the extra amounts being allowed for any wastage in the mixing.

The petre was refined or cleansed of impurities by boiling in a large copper vessel with as much water as would cover it. When it dissolved, the powder-makers poured the liquid into large tubs called settlers and left it to deposit the saltpetre, which settled to the bottom. The makers considered it unnecessary to repeat this operation, as much of the impurities had been skimmed off during the boiling. As soon as the solution was cool, it was pumped into the filtering trough and piped into filter bags, made from closely woven canvas. The liquid filtered through these bags into copper pans, where the liquid was left for twenty-four hours to form crystals.

When the crystals had formed, the liquid was drained off and the crystals were purified by adding more water, filtering again, and leaving them to dry. The purified saltpetre was melted in a copper pan to evaporate the water from the crystals. As soon as the saltpetre cooled and again started to form crystals, it was ladled into copper pans where it solidified.

At the Royal Laboratory, Woolwich, more care was taken with the saltpetre refining, the original process being repeated a number of times and the liquid being filtered through flannel bags.

The saltpetre, sulphur and charcoal were finely pulverised and then the correct proportions were put into a large chest, known as a mingling tub. The mixture was moistened and then blended ready for the powder mill.

Powder-makers were, by law, not allowed to work more than 42 lb of mixture in the process of milling, as accidents occurred when the stones came into contact with the bed and made a spark which ignited the powder. This was liable to happen only when the process was nearly finished, as the dampness of the mixture during the early stages would eliminate any risk.

The machinery of the powder mill, usually driven by water or 'horse' power, consisted of large cylindrical stones revolving in a trough with sloping sides. When the composition was sufficiently ground it was sent to the corning house for the next process.

At the corning house, the powder was granulated. This was done by using sieves with a base either of parchment or of bullock hide and perforated with holes of about a tenth of an inch in diameter. About twenty of these sieves were fixed in a frame, one above the other, and the mixture was placed in the top one. On the mixture was placed a circular runner of hard wood and the frame, fitted to an eccentric axis, was put into motion. The wood runners acted upon the powder and forced the mixture through in grains.

The grains were then glazed by placing about 200 lb of powder in a barrel and revolving it at a speed of forty revolutions a minute. This operation gave the powder a high or dull gloss, depending on the requirements, less time giving a dull appearance to the grains. The main object of glazing was to round the grains and give them a better appearance. Gunpowder intended for the African market was always made as bright as silver by adding black lead during the glazing as this appealed to the natives. This 'trade' powder was extremely inferior, as were the 'trade' guns in which it was used, and as a result powder for white settlers in South Africa was normally dull.

The powder was then laid out about $1\frac{1}{2}$ inches thick on canvas stretched on wooden frames and placed in racks in the drying room. In some mills this room was heated by steam passing through pipes but in others a stove or oven was used. The powder was taken, when dry, to the dusting house where the different-size grains were separated with sieves fitted in a frame. This process removed the dust from the powder. In the packing room the powder was packed into barrels, half barrels, quarter barrels, or loose in paper containers, according to the quality of the powder. The barrels were then sent to the Grand Magazine at Purfleet where the powder was examined and proved by officers of the Ordnance.

The instrument used for proving powder consisted of a metal socket, about the size of a lead pistol ball 0·65 of an inch in diameter, fixed vertically in a sturdy frame, with the muzzle pressed against a graduated bar fitted in a groove. The socket was loaded with a weighed amount of powder and then fired. The strength of the powder could then be gauged by the position reached by the bar.

When, through the passage of time or because of some accident, gunpowder became unfit for service, the saltpetre, being the most valuable part, was extracted and re-mixed with fresh sulphur and charcoal. Wherever gunpowder was stored, the barrels were never allowed to be left in the same position for any length of time because the heavier ingredient would sink to the bottom of the barrel and separate from the mixture. Every possible opportunity was taken to air the powder and to move the barrels about. Before cartridges were filled the powder was well stirred so that the power of each cartridge was constant.

In 1799, Sir William Congreve and General Sir Thomas Blomefield, the Inspector and Superintendent of the Royal Foundry at Woolwich, were ordered to inspect the guns and powder of the Fleet. They found, amongst other things, that there were only four barrels of well-prepared gunpowder in the entire fleet at Plymouth.

Because of the difficulty of stirring and mixing the powder at sea, there was often criticism of British powder, the shots of the French being said to outrange the British. However, this was probably due to the differing methods of gunnery in the opposing fleets at the time. The French, in particular, always aimed to destroy the rigging of a ship and therefore were more used to elevating their guns than the British, who usually fired at the hull.

CARTRIDGES
The first cartridges were paper bags filled with the correct charge of powder. By 1800 flannel had replaced paper and it was in turn superseded by serge a few years later. To make the cartridges, the charge of powder was measured out and was put into the bag which had three strands of worsted string sewn round the mouth. The worsted string was then twisted three times round the bag—at the neck, in the middle and at the bottom. This was done to keep the charge in its correct shape and to give the bag added strength. The process was known as hooping.

The cartridges used in the Royal Navy prior to 1800 were made of paper, soaked in a substance to prevent them from catching fire should any part be left in the barrel of the gun. It was still, however, very necessary to sponge out the gun and to draw out any piece of cartridge that might have remained in the bore, otherwise 'great mischief' would probably have ensued. The bottoms of the cartridge were found to remain more in some guns than others, and this was due to the drilling of the vents. It was also noted that guns primed with tubes were more liable to this than guns primed with loose powder.

One of the answers to this problem may have been that some of the powder slipped behind the cartridge when poured into the vent and so blew out any piece of the bag that might have remained when the gun was discharged.

Fixed ammunition, as it was called, was made by forcing a wooden bottom (*Sabot*) into the bag of the cartridge until it was firmly seated on the powder. The bag was then tied round the two grooves with worsted string and put into another bag which was choked in the grooves and fastened with catgut.

The projectile

There was a large variety of shot used in the pieces, both on land and at sea, during this period. Each type of projectile had its own special purpose, whether it was the destruction of walls and fortifications or setting fire to towns.

The earliest form of projectile was an arrow or iron dart, bound in leather or cloth so as to fit tightly into the bore. Although of little value, these projectiles continued in use even though stone shot had proved to be infinitely superior. The next step from stone was a cast-iron ball which gave better accuracy and a longer range.

Although these round projectiles were ideal for the walls of towns and opposing ships, they were less effective on troops, unless in a large, tight formation. For use against troops various other types of projectile were invented which would shower the opposing army with small particles of lead or iron. The first attempt at this form of shot was the loading of the guns with nails, flints, bolts, stones, in fact any small pieces of material that would be able to kill when showered on a formation of men. From this stage, these types of projectile became more sophisticated and specially manufactured. Canister shot, consisting of a case filled with small projectiles, was used as early as 1453, and the other types of shot were developed along these lines.

Stone cannon balls were liable to break when used against the walls of a town, and so the stones were hooped with bands of iron. This proved to be more effective than the stone alone. Iron shot was used in England in 1350, and lead shot as early as 1346. The accounts of John de Sleaford, Clerk of the King's Privy Wardrobe, show that between 1372 and 1374 men were employed at the Tower of London making lead 'pelottes'. The high cost of manufacture and the difficulties encountered in the casting delayed the general introduction of iron shot until the latter half of the fifteenth century.

In 1413 the British were still using stone shot and in some cases even arrows, and in 1418 Henry V ordered the Clerk of the Works of his Ordnance to procure suitable workmen to make 7,000 'stones' for guns of differing types in the quarries at Maidstone in Kent.

All types of metal were tried as cannon balls. The Venetians at the battle of Taro in 1491 used iron, bronze and lead cannon balls. Experiments were made with stone cannon balls covered in lead. By the end of the sixteenth century, iron balls were the only type of round shot in use.

At sea other types of projectile were used in the seventeenth and eighteenth centuries. These included bar shot and chain shot, which consisted of two cannon balls joined by a bar or a chain and were of great use against the rigging of an opposing ship.

ROUND SHOT

The most common projectile during the eighteenth and early nineteenth centuries was the round shot which was cast in iron to a certain weight and size intended for a certain gun. The main advantage of solid shot was its range, which was quite considerably further than that of any other form of projectile.

Round shot was most effective at a low trajectory so that it arrived at its target with high velocity and so was more destructive. Another advantage of firing at a low trajectory was that the shot tended to bounce after the first impact or 'graze' and the effect on opposing troops in formation was devastating, as many as 20–30 men being killed with one round shot.

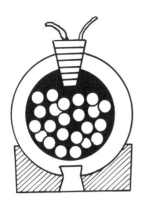

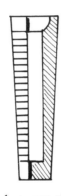

Another advantage of round shot was that it could be heated and, with due care, fired from guns. In 1575 the King of Poland successfully used red-hot shot in his siege of Danzig; the British used it with great success at the siege of Gibraltar between 1779 and 1783, and they had already experimented with it at Gibraltar in 1771. The main use of red-hot shot from a shore battery was against ships in the hope that the shot would burn its way through the decks of the ship and ignite the magazine, or even burn its way through the hull. The experiments carried out at Gibraltar in 1771 consisted of three tests.

(i) A red-hot 24 lb shot was placed between two pieces of fir timber, twelve inches square and hollowed out to take the shot. The timber soon caught fire and continued to burn for six hours until the wood was entirely destroyed.

13 Spherical case shot and round shot fitted with 'sabot'. Beech-wood fuses

(ii) A red-hot 24 lb shot was laid on the ground in the open air for four minutes and was then plunged three times into cold water and thrown amongst some cordage. In seven minutes the cordage broke into flame, and although an 'extinguishing engine' played water on the shot and cordage for two minutes, the cordage was alight again in fifty minutes.

(iii) A 32 lb red-hot shot, having been in the open air for four minutes, was immersed three times in cold water and laid between two pieces of oak about twelve inches thick. The oak was fresh and still full of sap, but as soon as the shot was placed on it, it began to smoke; in four hours there was a considerable fire,

25

in eight hours the wood fell apart and it was finally destroyed twelve hours after the start of the experiment. Twenty-one hours after the shot was removed from the fire, it was still too hot to touch.

The diameter of iron shot when heated increased by only about one-sixteenth, and so the guns' windage was still sufficient to allow the uninterrupted passage of the shot along the bore. When firing red-hot shot, wads were normally made from turf and placed between the powder and the shot to avoid any possible accident while loading the piece, but ordinary wads could equally be used provided they were dampened or moistened beforehand and provided they fitted the bore tightly. This moistening of the wad was said by Captain Robert Lawson, author of 'Artillery Memor-

andums relative to the Royal Navy' (1782), not to be absolutely necessary but done as a 'prudent precaution for general service especially when firing to windward'.

One great advantage that round shot possessed was that it was capable of being fired back at the original firer. A standard table found in a number of officers' personal notebooks of the period gives the various diameters of British and French shot (see Table 1). If, for example, the diameter of a French shot was found to be 6·64 inches, this measurement would be looked up in the column of French measurements and found to be a 36 lb shot. Next the measurement 6·64 would be searched for in the column of English measurements—the nearest measurements being 6·63 or 6·68. As 6·63 corresponds to a

Table 1

DIAMETER OF IRON SHOT, ENGLISH AND FRENCH, EXPRESSED IN INCHES

Weight (lb)	Diameter (English)	Diameter (French)	Weight (lb)	Diameter (English)	Diameter (French)
1	1.92	2.01	26	5.69	5.96
2	2.42	2.53	27	5.76	6.04
3	2.77	2.90	28	5.83	6.11
4	3.05	3.19	29	5.90	6.18
5	3.28	3.44	30	5.97	6.25
6	3.49	3.66	31	6.04	6.32
7	3.67	3.85	32	6.10	6.39
8	3.84	4.02	33	6.16	6.45
9	4.00	4.18	34	6.23	6.52
10	4.14	4.33	35	6.29	6.58
11	4.27	4.47	36	6.35	6.64
12	4.40	4.60	37	6.40	6.70
13	4.52	4.73	38	6.46	6.76
14	4.63	4.85	39	6.52	6.82
15	4.78	4.96	40	6.57	6.88
16	4.84	5.07	41	6.63	6.94
17	4.94	5.17	42	6.68	6.99
18	5.04	5.27	43	6.73	7.05
19	5.13	5.37	44	6.78	7.10
20	5.22	5.46	45	6.84	7.16
21	5.30	5.55	46	6.89	7.24
22	5.38	5.64	47	6.94	7.26
23	5.40	5.72	48	6.98	7.31
24	5.54	5.80	49	7.03	7.36
25	5.63	5.88	50	7.08	7.42

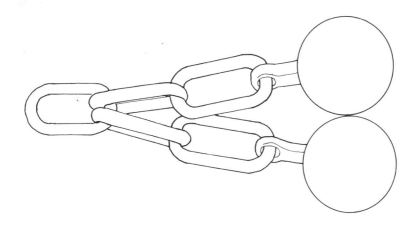

41 pdr, which did not exist, 6·68 would be selected as corresponding to a 42 lb shot. From this it could be established that the English could use a French 36 pdr shot in a 42 pdr gun.

DOUBLE-HEADED SHOT

This description takes in bar shot and chain shot which were mainly used by the Royal Navy. Before 1782 there was a large variety of double-headed and bar shot in use, some with two cannon balls or two half-balls connected by a bar and others with sliding bars that extended in flight. There were also knife-blade shot, with four folding blades which flew open in flight, and shot with a grappling hook attached which was designed to carry a rope to another ship. All these types of shot were found to be too weakly made and the cast iron easily damaged when fired, so there soon evolved only one kind made of wrought iron. This consisted of two solid halves of a shot connected with a bar. It was about three times the diameter of the ball in length, and weighed one third more than round shot of the same calibre, i.e. a double-headed shot for a 24 pdr weighed 32 lb.

The chief use of bar shot, or dismantling shot as it was sometimes called, was to destroy the rigging or sails of an enemy ship, but this

14   Chain shot

type of shot was not very highly regarded in the Royal Navy, owing to its inaccuracy.

During the American War of 1812, an action took place between the American ship *President* and HMS *Endymion* in which the American ship used dismantling shot. The British ship's sails were torn to pieces and the spars and rigging badly damaged. One of these shots, it was later stated, cut away twelve or fourteen cloths of the *Endymion's* fore-sail, stripping it almost entirely from the yard. American dismantling shot consisted of four or five iron bars, each about two feet long, fastened at one end to a ring, opening out like a star in flight.

CASE OR CANISTER SHOT

To make case shot, the shot was placed in a cylindrical tin case (the size of which depended on the calibre of gun for which the shot was intended) and sealed at the base with a wooden plug. This wooden plug, as mentioned on p. 23, was then forced into a cartridge bag and tied. The whole was then enclosed in another bag and tied with gut.

Case shot for howitzers had spherical wooden bottoms and the shot was placed in the tin at random, whereas case shot intended for

15 *Projectiles. Fixed ammunition, shells fitted to bags, canister shot of various calibres, shot charge cases, powder and small arms ammunition barrels*

the field artillery had the shot placed in regular tiers. This type of ammunition was first ordered for use on board ships of war in 1760, six rounds being allowed for each gun on the upper and quarter decks. Locks were supplied to fire the guns instead of common match.

Canister or case shot contained a considerable amount of shot, and in the Royal Navy was thought to be very useful for firing into the enemy's gunports to kill the opposing gunners when ships were alongside each other during a battle.

The balls for canister shot were iron, except for the shot for swivel guns which were made

of lead. Tables 2–4 give the number and weight of shot in each calibre of case shot on land and at sea, and Table 5 gives details of charge, elevation, and fuse required at various ranges.

SPHERICAL CASE SHOT

The forerunner of spherical case shot, or shrapnel as it became known, was the projectile invented by Master Gunner Samuel Zimmerman in 1573. Zimmerman's invention was a leaden cylinder with a fuse at the end next to the powder which was ignited on firing and exploded the shell in the air, scattering the bullets. The idea was never taken up and was soon forgotten.

The new type of shot was designed in 1784 by Lieutenant Henry Shrapnel, Royal Artillery. The spherical case shot consisted of a shell, much thinner than the common shell, filled with musket balls, with a fuse which would explode the shell at the correct moment.

The shells were always filled at the Arsenal and when they were sent out on service they were accompanied by fuses, the tangent scales of the guns being marked to correspond with the fuses and to ensure that the correct length was selected for the target.

During the manufacture of the spherical case shot, it was gauged both inside and out and the shoulder inside the fuse hole that kept the fuse in tightly was inspected. The shell was also gauged and then hammered on the outside to see if it was sound. Another test sometimes given to these shells was to put them in a barrel of water with a pipe securely fitted to the fuse hole and then to blow in some air from bellows and see if any bubbles rose from the shell. Conversely, water could be forced into the shell and then the outside inspected for any trace. The range of these shells was about 3,000 yards, and they were most effective when they burst about fifty yards from the target. The effect produced by these shells on troops in line was considerable.

The Ordnance were at first not interested in the new shell and it was not until 1803 that manufacture of it began. Shrapnel received a pension of £1,200 a year for life in 1814. He died in 1842 and some ten years afterwards the name of the shot was officially changed from Spherical case to Shrapnel at the request of his family.

It was later claimed by some that Shrapnel had only copied the design of Master Gunner Zimmerman and had not invented the shell that later bore his name. This is most unlikely, as the discovery of written reference to Zimmerman's shell was not made until 1852, ten years after Shrapnel's death.

GRAPE SHOT

The term grape shot stemmed from the appearance of the projectile, which was likened to a bunch of grapes. Quilted grape shot was designed to overcome the slight

Table 2

ENGLISH CASE SHOT FOR SEA SERVICE

Weight of gun (lb)	Weight of each shot (oz)	Number of shot in each case	Weight of case full lb oz
GUNS			
32	8	70	33 8
24	8	42	22 15
18	6	42	16 8
12	4	42	11 5
9	3	44	8 9
6	2	40	5 2
4	2	28	4 0
3	2	20	2 15
1	1¼	12	1 2¼
CARRONADES			
68	8	90	46 2
42	8	66	32 8
32	8	40	21 4
24	8	32	16 1
18	6	31	12 2
12	4	32	8 2

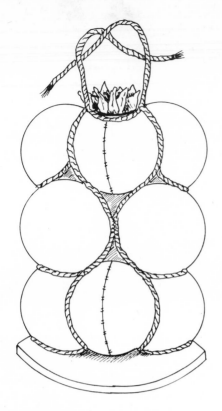

16 *Grape shot*

1800, the base and pillar of the grape shot for the land service were in wood, not iron.

For sea service, quilted grape shot was filled with nine balls only, irrespective of the calibre of the gun. When fired at an elevation, grape shot would range a considerable distance, but the effect would be lost at long range and so for this type of work langridge shot was preferred. Grape shot continued to be manufactured until 1868.

Grape shot was not fired from brass ordnance because of the damage it might do to the bore.

LANGRIDGE SHOT

Langridge shot consisted of a number of cast-iron bars usually enclosed in the same type of tin case as used in case shot and fixed to the charge. Experiments showed that these

defect of case shot which was that it sometimes arrived at the target still in one piece so that its effect was lost. Grape shot consisted of a number of iron balls arranged around a central iron column in a canvas bag. The column was attached to a base and the bag tied at the neck with the cords tied around the shot to give the appearance of quilting. The shot was then painted to preserve it.

In the first half of the nineteenth century, this type of grape shot was replaced by 'Mr Caffin's substitute' which consisted of three circular plates of cast iron with six holes in each and a fourth plate, of wrought iron, with three holes. The shot was placed between these plates, which were fastened together by an iron pin running up the centre and screwed tight at the top and bottom with a nut. Before about

Table 3
ENGLISH CASE SHOT FOR LAND SERVICE

Weight of gun (lb)	Weight of each shot (oz)	No. of shot in each case
42	8	85
32	8	66
24	8	46
18	6	46
12	4	46
9	3	44
6	2	40
4	3	28
3	$1\frac{1}{2}$	31
2	$1\frac{1}{2}$	20
1	2	10

Table 4
CASE SHOT FOR HOWITZERS

Weight of howitzer (lb)	Weight of each shot (oz)	No. of shot in case
$4\frac{2}{5}$	2	55
$5\frac{1}{2}$	2	100
8	2	258
8	8	72
10	8	170

Table 5
CHARGE, ELEVATION AND FUSE FOR SPHERICAL CASE SHOT

Range (yds)	charge (lb)	elevation (degrees)	fuse (inches)	charge (lb)	elevation (degrees)	fuse (inches)
		24 pounder			18 pounder	
650	6	$1\frac{1}{4}$	0·2	$4\frac{1}{2}$	$1\frac{1}{4}$	0·2
900	6	$1\frac{3}{4}$	0·35	$4\frac{1}{2}$	2	0·4
1,100	6	$2\frac{1}{2}$	0·5	$4\frac{1}{2}$	3	0·573
		Medium 12 pounder			Light 12 pounder	
650	4	$1\frac{1}{4}$	0·2	3	$\frac{1}{2}$	0·275
900	4	$1\frac{3}{4}$	0·4	3	2	0·475
1,100	4	$2\frac{1}{2}$	0·515	3	$3\frac{1}{4}$	0·65
		9 pounder			Long 6 pounder	
650	3	$1\frac{1}{4}$	0·225	2	$1\frac{1}{2}$	0·225
900	3	$1\frac{3}{4}$	0·4	2	$1\frac{3}{4}$	0·45
1,100	3	$2\frac{1}{2}$	0·6	2	$2\frac{3}{4}$	0·65
		Light 6 pounder			Long 3 pounder	
650	$1\frac{1}{2}$	$1\frac{1}{2}$	0·3	1	$1\frac{3}{4}$	0·3
900	$1\frac{1}{2}$	2	0·5	1	$2\frac{1}{2}$	0·6
1,100	$1\frac{1}{2}$	$2\frac{3}{4}$	0·65	—	—	—
		Light 3 pounder			68 pounder carronade	
650	$\frac{3}{4}$	2	0·3	4	$2\frac{1}{2}$	0·4
900	$\frac{3}{4}$	$3\frac{1}{4}$	0·6	4	$3\frac{1}{2}$	0·6
1,100	—	—	—	4	5	0·85
		24 pounder carronade			Heavy $5\frac{1}{2}$ inch howitzer	
650	2	—	0·375	2	$3\frac{1}{4}$	0·45
900	2	—	0·6	2	$4\frac{3}{4}$	0·65
1,100	—	—	—	2	$6\frac{1}{4}$	0·65
1,200	—	—	—	2	$7\frac{1}{2}$	0·71
		Light $5\frac{1}{2}$ inch howitzer				
650	1	$5\frac{3}{4}$	0·6			
900	1	8	0·9			
1,100	1	10	1·33			

Table 6
GRAPE SHOT FOR SEA AND LAND SERVICE

Weight of gun (lb)	Weight of each shot (lb oz)	Total weight of grape (lb oz)
42	4 0	46 6
32	3 0	34 1
24	2 0	25 5
18	1 8	19 $15\frac{1}{2}$
12	1 0	10 15
9	0 13	7 6
6	0 8	5 $8\frac{1}{2}$
4	0 6	3 $14\frac{1}{2}$
3	0 4	2 $10\frac{1}{2}$
$\frac{1}{2}$	0 $\frac{3}{4}$ (lead)	0 $8\frac{3}{4}$

bars were very effective at sea in the destruction of rigging and sails of a ship.

A large variety of other shot based on this design either for burning or cutting rigging had been suggested at different times, but they usually proved too weak to stand the explosion, or were not as efficient as langridge shot.

CARCASSES

The main object of carcasses was to set fire to an enemy's town, stores or ship. The composition of the carcass was such that it was almost impossible to extinguish. The earlier carcasses were made from canvas and then covered in tar but these were replaced by cast-iron cases in which were 3–5 holes. The composition was made by melting turpentine and tallow together in a pot which stood in a pot of boiling oil. This increased the heat yet diminished the danger of any accident. The rest of the composition, consisting of various measures of saltpetre, sulphur, resin and antimony, was then added and mixed in.

17 Projectiles. Common shell and shot gauges

A funnel was put in the fuse hole and the mixture was rammed in with a drift. The fuse or fire holes were then filled with fuse powder. Carcasses were fired from mortars and howitzers only. In 1808, for the Scheldt expedition, explosive carcasses were prepared under the direction of Sir William Congreve. These were fitted inside with a flintlock which was activated by a clockwork mechanism which could be set for any time under $2\frac{1}{2}$ hours.

COMMON SHELL

Common shell consisted of a hollow round shot, filled with powder and suitably fused so that the powder inside would explode and fragment the case. Before 1700 the method of firing these shells was either to face the fuse on the charge so that it would ignite when the piece was discharged or to face the fuse upwards and light it before the gun was fired. In 1700, however, it was discovered that an unlit shell with the fuse upwards would be ignited, when the gun fired, of its own accord by the flash of the powder. The early forms of shell had the case thicker at the bottom than at the fuse

18 *Parachute flare enclosed in metal case*

end with the idea that the shell would arrive with the fuse uppermost and not likely to be extinguished, which it would be if the fuse was buried in the ground. This theory proved false, and so later shells were made of uniform thickness.

HAND GRENADES

These were small cast-iron shells filled with powder, and up to the 1770s they were thrown by hand. They had a delayed fuse which was lit before throwing. The grenades used at sea were slightly larger than those for the land service and had an external diameter of 3·5 inches, weighing 3 lb 11 oz. The charge was 3 oz of powder. There was a cup attachment which could be fitted to a musket and was first designed for use in the woods of the West Indies. A blank charge of 6 drachms of powder would propel a grenade about 250 yards. The cup fitted on the barrel of the musket in the same way as the bayonet.

In the Royal Navy it was thought that a grenade cup attached to a swivel gun could be effective. There is, however, no strong evidence to show that the grenade-thrower ever came into widespread use in the Navy.

COMPOSITIONS PRODUCING LIGHT AND SMOKE

These compositions, which were often used on land, were found in a number of cases to be very useful in the sea service. The use of light balls was to illuminate a given area or target, while smoke balls were used for concealment and in certain cases to suffocate men working in confined spaces.

One type of light ball was made from mealed powder, saltpetre, and sulphur, which gave a strong bright flame. Another type of light ball was made of saltpetre, sulphur, resin and linseed oil. The shells used had a framework made of iron, oval in shape and with a hemispherical iron base. This frame was covered in cloth and filled with the mixture through three holes. These holes were then fused and the whole shell painted.

Smoke balls were composed of saltpetre, corned gunpowder and sea coal to which sometimes antimony was added. This mixture produced heavy dark smoke. Light and smoke balls were fired from either mortars or howitzers. For smoke balls, the cases were made from papier mâché and filled with the mixture. The case was then painted over with tar.

Light balls were considered useful in the

Navy for guard ships stationed in rivers, as the balls floated on water and lit up a considerable area. The only objection to them was that they burned only for about five minutes.

An example of the use of smoke is given by Captain Robert Lawson.

> The smoke composition might be applied in stratagems for deceiving an enemy—as for instance, suppose a ship nearly subdued by a superior force, a number of these balls being placed in any convenient part of the ship and set fire to would produce so violent a smoke as to induce the enemy to suppose the ship totally on fire; and of course endeavour to avoid it. This might produce a favourable opportunity for the inferior ship to make her escape.

ROCKETS

Rockets had been used as a weapon of war for many centuries, mainly by the Chinese and the Indians. Tippoo Sahib used rockets against the British at the siege of Seringapatam in 1799 with considerable success, and this prompted

19 Different types of rockets in use, with tools and rocket parachute flare

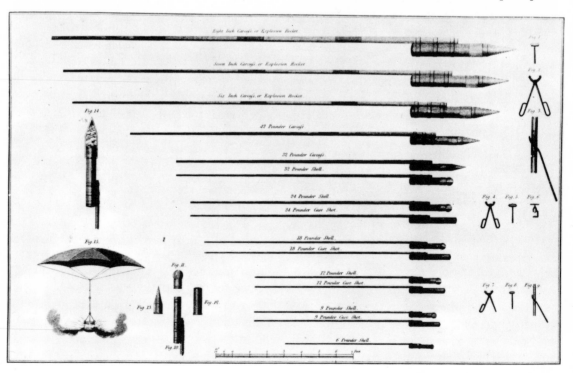

the Ordnance to apply to Woolwich for the

the Ordnance to apply to Woolwich for the names of suitable manufacturers of rockets. The enquiry was directed to the East India Company by the Royal Laboratory, but the Company replied that they were unable to help. It was owing to the efforts of William Congreve that rockets were introduced into the British service.

In 1805, Congreve persuaded the Government to allow him to try rockets in action. At the attack on Boulogne on 18 November, a gale arose which put half of his rocket-carrying boats out of action. In October 1806, however, eighteen ships were used in another attack in which two hundred 3 lb rockets were fired in half an hour. In 1807, 40,000 rockets were used in the attack on Copenhagen. On land, rockets were found to be erratic. This is not surprising, considering that in many cases they were fired along the ground.

In 1814, Rocket Troop was formed and Colonel Sir William Congreve produced his instructions for their use in a book entitled *The details of the Rocket System*. The number of men and horses in a rocket troop was the

20 Rockets fired from launches by Marine Artillerymen

same as in a horse artillery troop.

Rockets were divided into heavy, medium and light. The heavy rockets were 8, 7, and 6 inches (size denoted by diameter of head). They were designed as explosives or for setting fire to targets. The medium rockets were the 42, 32 and 24 pdrs. The heads of these rockets were made interchangeable so that any of them could be a carcass rocket, a shell rocket or a case shot rocket. The light rockets were 18, 12, 9 and 6 pdr. At the time of writing his book, Congreve only speculated about the interchangeable heads, but in 1814 they were experimented with.

The sticks of the larger rockets were made in sections and joined with iron ferrules, while those of the smaller sizes were in one piece. The 42 and 32 pdrs were principally used in bombardments for carrying carcasses into the enemy fortification, but the 32 pdr could also be armed with a shell. The 18 pdr rockets were equipped with a 9 pdr shell, the 12 pdr with a

35

21 *The use of rockets in fireships and for coastal bombardment. Lower left and right show the method of adapting a ship to fire rockets*

6 pdr shell, and the 9 pdr with a grenade. Special elliptical shells were made for use with rockets as these were found to increase the velocity in flight. Explosive rockets were fused and ignited by the vent when the rocket was fired. The fuse, as with a shell fuse, could be cut to suitable lengths.

Rocket sticks for the land service (used by Rocket troop) were made in sections, while those intended for use at sea were made in one piece. There were also signal rockets which weighed 1 lb or $\frac{1}{2}$ lb and were composed of saltpetre, sulphur and dogwood charcoal. A 6 pdr rocket had a stick 7 feet long and a 12 pdr rocket a stick 9 feet long. The cases were originally made of paper, but in 1806 this was replaced with a sheet iron case and the sticks were shortened to give better balance in flight.

Table 7

TYPES OF ROCKETS IN SERVICE, LAND AND SEA, 1814

Nature	Length of stick (approx)
8 in	24 ft
7 in	22 ft
6 in	21 ft
42 pdr	17 ft
32 pdr	15 ft
24 pdr	13 ft
18 pdr	11 ft
12 pdr	9 ft 6 in
9 pdr	9 ft 6 in
6 pdr	8 ft

In 1817, Sir William Congreve established a factory for the manufacture of rockets and in 1815 he had improved on his previous design by placing the stick centrally in the base instead of to one side.

Rockets were fired from rocket cars, which were carriages suitably fitted out to take the rockets. There were two types, the heavy and the light. The heavy could carry forty rounds of 24 pdr rockets armed with coehorn shells and the light car could carry sixty 12 pdr or fifty 18 pdr rockets. The sticks were carried in half lengths in a box on the back of the carriage. Rockets were also fired from frames of differing design, as well as along the ground and from earth or stone banks.

In the sea service, rockets were fired from small boats with a frame suitably positioned on a forward mast. The sails were kept constantly wet during the firing to avoid the risk of fire. Rockets were also put aboard fireships and William Congreve stated in his book that he experimented with them aboard fireships in Basque Roads against the French fleet with great success, as the rockets going off prevented any enemy boats getting a line on board and towing the burning vessel away from their ships.

Congreve fitted rocket chutes to the defence ship *Galgo* and the sloop *Erebus* and stated in his book that all the gun brigs on the Boulogne station during Commodore Owen's command were fitted with rockets in frames.

4 IGNITERS, FUSES, CANNON LOCKS AND GUN EQUIPMENT

The earliest method of igniting the charge in a cannon was by plunging a hot iron into the powder over the vent, but by 1500 this method had been superseded by the portfire and the linstock. The linstock held slow match which was kept burning by the gun when in action to provide fire for the portfire which was used to fire the piece. The linstock had a wooden shaft and a spear head each side of which was a branch; the branches were sometimes decorated with serpents' heads. The match was fitted into the jaws at the end of one of the branches and coiled round the shaft of the linstock. The butt of the linstock was also fitted with a sharp pointed shoe so that it could be driven into the ground by the guns when in action.

The portfire-holder carried the portfire which was used to ignite the gun. It was lit from the linstock and when no longer required was cut off with the portfire-cutter which was mounted on the trail of the gun carriage.

Slow match, used on the linstock, was made from loosely woven strands of hemp, boiled in old wine dregs or in a solution of wood ash or saltpetre. The hemp was then bound with an outer layer of strands. Slow match burned at a rate of about a foot in three hours. Portfires were made in the following way. A sheet of paper was cut into a square. This was then pasted and rolled onto the mould or former. The bottom of the case was turned in and the mixture put inside and driven tight with a rammer. (The mixture consisted of saltpetre, sulphur, and mealed powder. This was thoroughly mixed and sieved before being put in the case.) Portfires were usually about sixteen inches long and burned for between twelve and fifteen minutes.

The corrosive properties of gunpowder soon led to the introduction of a new method of containing the priming charge for the cartridge. This took the form of a length of quick match enclosed in a tin case which could be inserted into the vent of the gun. This tin gave way to the common paper case or tube and the quill tube, which was made from a goose feather. The quick match, which up until about 1800 was used in the quill and paper tubes, was made of 3–6 cotton strands which were doubled and boiled in a copper pot. Saltpetre was added before the water boiled, and the mixture was boiled until nearly evaporated when the spirits of wine were added and the mixture boiled again. Mealed powder was then sprinkled in and left to cool for about six hours.

The mixture was then placed in another pan of mealed powder for six hours before being rolled on a reel.

In 1800, the match was replaced with a composition of mealed powder, moistened with spirits of wine. Paper tubes were made from small arms cartridge paper, cut into pieces of a rectangular shape, $5\frac{1}{2}$ x 2 inches. These were then rolled on a former and stuck; when dry, they were cut into lengths of $1\frac{3}{4}$ inches. Having been filled, they were capped with a small piece of cartridge paper to keep the mixture in.

Quill tubes were made from goose feathers which were cut and gauged on the rounding board before being trimmed and scrapped. The quill was then cut to a length of about $2\frac{1}{2}$ inches, and slit at the top. The top portions were then bound in worsted, the strands passing through the slit. The quill tubes were then filled with the mixture, making sure that it was completely solid. A thin wire was passed through the composition and a small amount of paste of the same mixture was put in. This was called 'priming the tube'. The quill was then sealed with a small piece of cartridge paper.

Quills and paper tubes were nearly always used at sea as they could be made up on board ship at any time. It was estimated that one man could make 200 quills in a day. The implements needed for making quill tubes were an iron gauge, 0.2 inches in diameter, a pair of scissors, a rounding board, a pair of shears, a nipping press, a probe, a knife with seven prongs, and a needle and worsted.

Although flintlocks had been used on military muskets and pistols since the early 1700s they were not generally used on artillery on land or at sea during this period. There are examples of orders in existence introducing cannon locks for the sea service, and in a number of cases these were used. It was not until 1840 that the percussion lock was finally

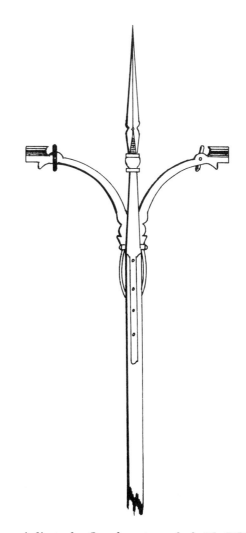

22 A linstock. One langet marked 6th RGB (*Royal Garrison Battalion*) 1802–6. Tower of London collection no 7/1149

adopted for general issue, this in turn being replaced by a friction tube in 1853.

The Royal Navy had introduced cannon locks as early as 1755 by Admiralty Order of 21 October of that year. In 1760, when case shot was first supplied to the Navy, locks were also issued for the guns of the upper and quarter decks. Major-General Sir Howard Douglas, in his *Treatise on Navy Gunnery* (1820), attributes the introduction of cannon

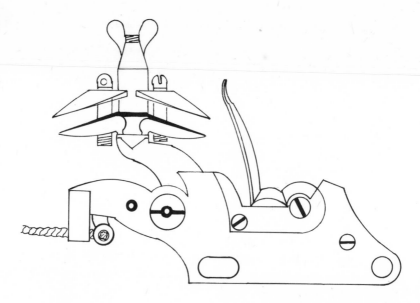

23 *Douglas pattern cannon lock*

locks to his father, Captain Sir Charles Douglas, who provided them at his own expense on board the ship *Duke*.

Sir Howard Douglas was responsible for the invention of his improved lock, or the Douglas lock, in 1817. He was engaged in correspondence with the Board of the Admiralty and Lieutenant-Colonel Sir Alexander Dickson KCB of the Royal Horse Artillery concerning the new lock. In a letter of 10 September 1817, the Admiralty replied to Sir Howard Douglas:

> I have the honour by their commands to acquaint you that the Lord Commissioners of the Admiralty have transmitted a report from Admiral Sir Edward Thornbrough on the subject of the trial which has been made with the locks sent to Portsmouth; by which it appears that the locks are considered to be a very great improvement on those present in use.

A further letter dated 16 January 1818 added the following comments:

> . . . it has been decided that the provision of locks for Sea Service Ordnance now in use should be discontinued, and those of your invention gradually introduced into the service.

The introduction, however, as with all new improvements in the services, took some considerable time to be implemented.

Lieutenant-Colonel Dickson, of the Royal Horse Artillery, while praising the invention, was sceptical, adding in his letter of 20 April 1818 that although the locks would diminish the expenditure of match and portfires,

> I would not on that account propose the conveyance of a less quantity of those articles than now is practice for in close action the lock could not be depended upon and in bad weather it could not be used.

The Royal Artillery had, however, introduced the locks experimentally for the larger calibre garrison and siege guns, not making them general issue until well into the 1830s.

The standard cannon lock was a brass-framed flintlock box lock, the pan connecting with the bottom plate by a channel. At each end there was a screw and wing nut to secure the lock to the vent field on the breech of the piece. The pan was primed with powder, the frizzen shut and at the same time the hammer holding the flint was cocked. A lanyard was connected to the firing lever at the back of the lock which, when the lanyard was pulled, released the hammer whose flint sparked on the steel of the frizzen, throwing it forward and igniting the powder which in turn ignited the main charge.

The improved Douglas lock was of similar design except that there was a double set of jaws, holding the flint, on the hammer. These jaws were reversible by simply undoing the wing nut at the top and swivelling the new flint into place. One of the reasons stated by Sir Douglas for its introduction was that he had observed men using linstocks and match when the flint of a normal cannon lock was no longer sparking, rather than, in the heat of battle, stop firing to change the flint. A further advantage was that fewer loose flints were needed.

Another way of firing artillery was by means

of a cannon igniter, two examples of which are in the Armouries of the Tower of London. Their general appearance is of a cut-down and converted cavalry pistol with a long copper tube fitted to the short barrel. The copper tube is held in different ways on the two specimens in the Tower. The first method is a screw fitting and the other is a bayonet fastening. The igniter was loaded as a normal pistol, but without the ball and the copper tube attached. When the order came to fire, the end of the tube was placed over the vent and the trigger of the pistol was pulled. The action of the flintlock ignited the charge in the barrel of the pistol which sent a flash down the copper tube which passed down the vent and ignited the main charge. The drawbacks of

24 *Rocket muskets and cannon igniters*

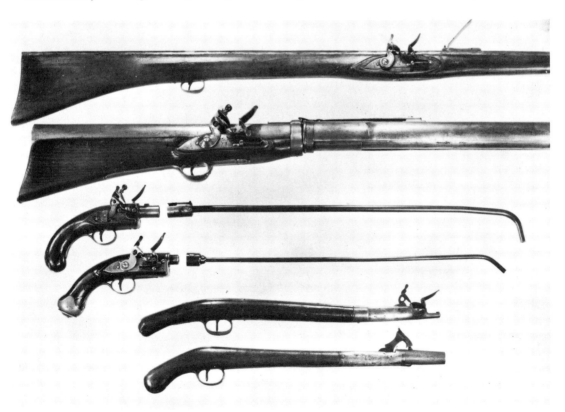

this method of ignition are obvious, the main ones being the time taken to reload the pistol and, again, its dependability. It is doubtful whether these methods were ever introduced on any scale. One of the examples in the Tower of London has engraved on it: 'Pattern flash pistol proposed for Royal Horse Artillery 1816' (Tower of London catalogue number 12/911).

There are also examples of 'flash pistols' of a different construction in both flint and percussion. They have long grips and are obviously specially made for igniting.

The use of explosive shot led to the invention of the fuse. As has been stated in Chapter 3, the early fuses were made from match which was put into the vent of the shot and either faced towards the charge or towards the muzzle depending on the method of ignition employed. By 1700 it was found that the fuses lit themselves and there was no need for them to be ignited prior to the piece being fired.

By the early 1700s, quick match had been replaced by an iron tube filled with a composition of saltpetre, sulphur and mealed powder. The iron tube was driven into the vent in the shell. By the 1750s, it was found that beechwood was not only cheaper but easier to make and handle than the iron tube and beechwood fuses were introduced. The wood was shaped and drilled before being allowed to season for a few years. The composition was then driven in, together with two strands of quick match. The top was sealed with a piece of parchment.

The outside of the wood fuse was graduated with small cuts, each of which represented half a second. These could be cut off to leave the appropriate length of time when the distance of the opposing enemy had been ascertained. If the projectile was required to burst in the air over the target, great care had to be taken to make sure the fuse was cut to the correct length, but if the shell was to burst on the

ground the whole fuse could be left intact.

One disadvantage beechwood fuses had was that they were prone to shrink in hot climates which resulted in the fuse falling out of the shell in flight, or the composition falling out.

Percussion and concussion fuses are outside the scope of this book, though John Muller in his *Treatise of Artillery* (3rd edition, 1780) states that a certain Lieutenant Pirle, who was lost on board the *Dodington* on its way to the East Indies, had invented a concussion fuse, but, 'being too modest a man, had not the assurance to propose it to the master general of the ordnance, whereby the world was deprived of so useful an invention'. In 1846 Quartermaster Freeburn of the Royal Artillery invented the concussion fuse that exploded the shell on impact.

In order to handle, load, fire and clean a gun in action a large variety of tools or side arms was needed. A contemporary handwritten notebook of an officer cadet of the period gives the disposition of ammunition and stores on a 6 pdr carriage and is a good example of the type of equipment carried by the Artillery.

Right side box

Round shot fixed to cartridge	6
Case shot fixed to cartridge	6
Tube box with 80 tubes	2
Portfires	6
Portfire stick	1

Left side box

Round shot fixed to cartridge	6
Case shot fixed to cartridge	6
Portfires	6
Portfire stick	1
Sponge tacks	100
Punches for vents	2
Set of wires	1
Spikes common	1
Spikes spring	1
Couples of chain traces	2
Spare washer and linch pin	1 each
Tarred Marlin . . . skin	1

Locker under the gun

Round and case shot unfixed for emergency	6 & 4
Claw hammer	1

Lashing rope of the crane	2 fathoms
Fore limber box	
Round shot fixed to cartridge	8
Case shot fixed to cartridge	8
Flannel cartridges 3 oz	14
6 oz	11
8 oz	8
Slow match	1
Hind limber box	
Round shot fixed to cartridge	8
Case shot fixed to cartridge	8
Flannel cartridges 4 oz	14
5 oz	10
7 oz	10
Portfires	24
Sheep skin	1
Cartouches of leather laid on the	
limber box	2
Locker under the limber	
Rope	45 fathoms
Blocks strapped with iron	3
Behind the axle tree	
Felling axe	1
Hand-bill	1
Pickaxe	1
Spades	2

A bucket was carried under the breast-transom set of men's harness under the gun carriage. A set of drag ropes on the front of the fore limber box and one set under the gun or limber.

Side arms	
Right side of the gun	
Sponges	2
Ladle	1
Left side of the gun	
Straight traversing handspike	1
Crooked handspike	1
Wad hook	1
Between the cheeks	
Fork lever	2
Linstock (on left cheek)	1
Length of tackle and lashing rope	
Tackle long	30 fathoms
Tackle short	15 fathoms

There was a large amount of ropes all cut to length for lashing the various implements. If only twenty-four rounds were carried, they were placed in the limber and the empty side boxes carried in the waggon.

Other most important pieces of the gunner's equipment were the quadrant and the tangent scale. The quadrant measured the tangent elevation by means of a setsquare, suitably calibrated, put into the muzzle of the piece. A plumb-line fixed to the apex then gave the reading. The tangent scale was fitted to a hole in the base ring of the piece and provided the angle of elevation of the target. Another table which helped the accurate laying of the guns was the quarter sights. These were engraved lines in the base ring of the piece. Mortars were laid by a plumb-line along the line of metal for direction, and if elevation was required a plumb-line was held over marks on the trunnions which provided the angle of elevation. Clinometers were also used to get an accurate reading of the angle of the carriage. Set tables were supplied giving various ranges of pieces at certain elevations.

In the Royal Navy, less equipment was needed, as the guns were fixed in position. On board ship each gun had beside it the rammers, sponges and wad hooks. There were also handspikes for elevating the guns. In the Navy a rope rammer was used, as it was easier to handle in confined spaces. In time of action, buckets, shot and cartridges were brought out.

5 THE CONSTRUCTION OF GUNS AND CARRIAGES

Guns

Brass and iron guns were cast in a mould consisting usually of three parts: the length of the gun, the cascable and the deadhead. The model was formed by taking a wooden spindle, which was sometimes part of a mast from an old ship, longer than the intended barrel. This spindle was placed over a trough with the ends resting on blocks which were about two feet from the ground. This spindle was then bound with plaited straw or rope strands to form the rough ground work before the composition was applied and the final shape of the barrel made. The composition was a mixture of either clay, brick dust and grease or loam, sand and horse dung.

This mixture was laid over the straw- or rope-covered spindle and the model was placed over a charcoal fire to dry and harden. This process was repeated until the model had been built up to the correct thickness. The final shape of the barrel was modelled at the last stage by holding a template cut to the shape of the barrel against the model which was slowly revolved.

The model was then washed over with 'tan-ashes' to prevent the mould from sticking to it. It was well dried over a charcoal fire and then pieces of wood were nailed on to form the trunnions, and any other decorations (such as dolphins and crests) were put on. The mixture used for making the decoration was composed of resin and beeswax. The vent field was in lead.

The cascable was formed on an iron spindle in the same way as the length of the gun and placed over an iron cup to give it added strength as it would bear the total weight of the metal during casting. The deadhead was made in the same way.

Once the model had been made, a mould was made from it, usually in three parts so that it could be easily removed from the model. The component parts of the mould were then banded together with iron reinforcing strips which served to keep the parts in place and to add strength to the clay during the cast.

The pit was filled with clay and the mould was put back with passages for the pour made from brick. These brick ducts were heated by charcoal to ensure that the metal would flow evenly during the cast. About 6–8 guns were cast at once using this method, with a small steel door in the troughs to prevent the molten metal from entering more than one

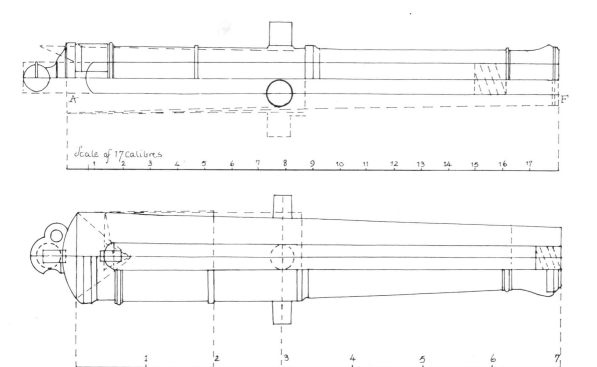

Scale of 17 calibres
1 2 3 4 5 6 7 8 9 10 11 12 13 14 15 16 17

1 2 3 4 5 6 7

gun at once. Any superfluous metal ran into a brick pit where it was allowed to cool, to be used again.

Three or four days after the cast, the barrels were withdrawn from the mould by a crane but, even so, it was still some time before they could be handled. Having been removed from the pit, the hoops, bars and clay were knocked off and the deadhead was removed. The centre of the bore of the gun was then accurately measured and marked. The guns were then taken to the boring and turning room.

The deadheads were remelted for the next cast. Copper possesses great strength and melts at 1,200°F; when it is combined with tin, which melts at 442°F, the compound is more fusible but less malleable, and so ideally suited for guns.

From 1750, sand moulds were also in use with the model made from copper instead of being made up as described above. The bore was originally cast into the barrel by putting a

25 The general construction of brass guns. The scale beneath shows the proportions of each part expressed in calibres. The calibre of a 6pdr being smaller than a 9pdr, the various parts are proportionately smaller

26 The general construction of iron guns. The scale beneath shows the proportions for the various parts, the size varying with calibre but the proportions remaining the same

clay core into the mould, but in 1739 Maritz, a gun-founder from Geneva, devised a method of boring a solid-cast barrel. From 1750 onwards nearly all guns were cast solid and bored afterwards.

In an Ordnance Minute of 9 August 1775, the Board stated that it would not receive any guns that were not bored from the solid. It appears that up until this date gun-founders had the option to quote for either bored from the solid or cast with the bore.

Casting of ordnance had to be carried out

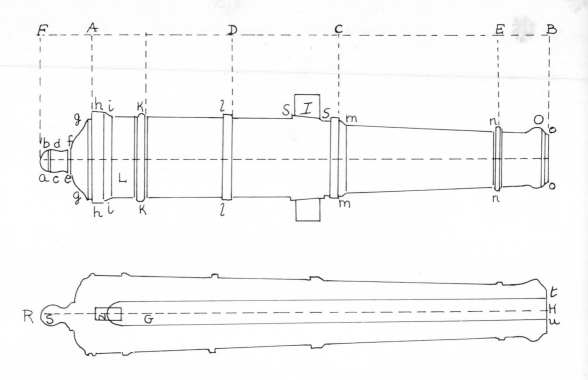

27 The names of the several parts of a gun. AB, length of the gun; AD, first reinforce; DC, second reinforce; CE, chase; EB, muzzle; FA, cascable; RH, axis of the piece; GH, bore; I, trunnions; L, vent field; N, vent; O, swell of the muzzle; ab, button astragal; ef, neck fillet; gg, breech ogee; hh, base ring; ii, first reinforce ring; mm, second reinforce ring; nn, muzzle astragal and fillets; oo, muzzle mouldings; ss, shoulder of the trunnions; tu, diameter of bore or calibre; S, button

with the utmost care, as any dampness in the mould could cause steam which might explode the mould and shower metal in all directions. There is an instance of this happening. In 1704 the largest brass gun foundry in England was in Windmill Hill, Moorfields, in the City of London, run by Matthew Bagley who was also a bell caster. In May 1716, a number of distinguished guests were invited to watch cannons being cast. Because of dampness in the mould, the molten brass exploded killing

Bagley, his wife and son, plus nine workmen. General Borgard, the chief ordnance officer from Woolwich, was seriously injured.

The death of Bagley deprived England of its one and only founder capable of casting brass guns, and the Government was forced to bring a founder named Schalch from France to supervise the casting of guns at Woolwich.

After casting, the barrels were taken to be bored and turned, both processes being carried out at the same time. The gun was hand-finished with files and the centre marked at the muzzle end and at the end of the button. The barrel was fitted to the machine which was run by water or 'horse' power.

The muzzle was locked in the chuck and collar with the button held firmly in a socket, which was attached by a wheel to the machine. The bar with the boring bit was laid on the bench in line with the mark on the muzzle of the gun. To ensure absolute straightness, three plumb-lines were lowered from a bar above to

28   A brass gun of the reign of George I spoiled
in the cast at the foundry at Moorfields in 1716
in consequence of the metal being allowed to run
into a damp mould

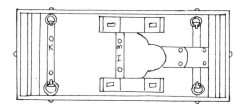

29   The names of the several parts of a land
service mortar bed. a, capsquares; b, eye bolts;
g, rings with bolts; d, upper and under bed bolts;
I, middle plates; k, end rivetting plates; m,
rivetting bolts; p, traversing bolts; q, key, chains
and staples; r, joint bolts

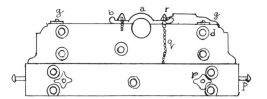

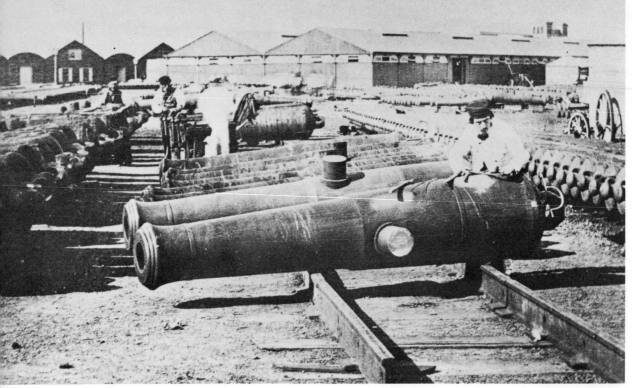

30 *Bushing vents of barrels at the Royal Arsenal. From a photograph of 1858*

check direction and every part was checked for tightness, as the least variation would spoil the barrel.

The machine turned the barrel and the boring bar was moved slowly forward by a man who sat at the far end of the bench turning the wheel. As the barrel revolved the bit dug into the metal and formed the bore. The bit varied in size according to the calibre of ordnance that was being bored.

Howitzers and mortars were bored in a slightly different way, a small bit being first used to form the chamber before the larger bit was used to form the bore.

The external surface of the barrels, having been roughly finished by hand, was turned on a lathe, with the exception of the part between the trunnions which it was impossible to turn.

The next process in the course of manufac-

ture was the drilling of the vent which connected the vent field, where the priming powder, quill or lock was placed, to the bore which carried the charge.

From about 1812 onwards, experiments were conducted into the bushing of vents with a copper plug. This was found to be a practical solution to oversized vents which occurred either from erosion or from much use. (Later, in 1844, iron bushes were used but never proved as satisfactory as copper and a return was made to the latter in 1855. After 1855, all newly manufactured guns were bushed prior to issue.)

Bushing was carried out by tapping the vent with about five or six threads and screwing in a copper cone which had a hole already drilled in the centre. If the cone projected into the bore, a cutter was used to make it flush with the internal diameter.

The piece was then sent to have the trunnions finished. The barrel was placed in a frame with the trunnions vertical and a cutter was then placed over the uppermost trunnion and turned with iron bars. Heavy weights were placed on the machine which kept the cutters tight on the trunnion and made sure they were cut accurately. Once this operation had been completed the hole for the tangent scale was bored and the parts of the gun which were unable to be turned on the lathe—i.e. between the trunnions and the parts near the vent field, sight and chase—were filed smooth by hand.

At this stage the gun was ornamented. Apart from guns which had dolphins, the ornamentation consisted of the cypher of the reigning monarch and the initials of the Master-General of the Ordnance. On iron guns and brass guns cast before about 1760, the cypher and initial were raised but for brass ordnance after that date they mostly appear to have been indented. The period covered by this book has only one reigning monarch, namely George III, but undoubtedly there were pieces from the reign of George II (1727–60) still in service. The position of the

cypher was usually on the first reinforce. Queen Anne had used the crown and the rose but the Georges had used their cyphers. For George II, III and IV, the numerals were sometimes incorporated. On certain pieces prior to 1750, the entire royal coat of arms was sometimes used.

The single initial of the Master-General of the Ordnance together with the coronet appropriate to his rank were placed on the chase. Howitzers were normally marked with the royal cypher on the chase and the initials of the Master-General of the Ordnance between the trunnions. After the abolition of the Board of Ordnance in 1855, the initial of the Master-General no longer appeared.

During the period 1790–1820 there were the following Masters-General of the Ordnance:

1784–95	Charles, Duke of Richmond
1795–1801	Charles, Marquis Cornwallis
1801–6	John, Earl of Chatham
1806–7	Francis, Earl of Moira
1807–10	John, Earl of Chatham
1810–18	Henry, Earl Mulgrave
1819–27	Arthur, Duke of Wellington

The name of the founder and the date were usually placed on the base ring in Roman figures. The weight in hundredweights, quarters and pounds, each separated by dots, was marked on the top of the breech. The broad arrow mark denoting government owner-ship was placed on the barrel, although there appears to have been no rule as to where. The quarter-sight scales were engraved on the base ring. Other marks, such as numbers on the trunnions, were normally manufacturer's serial numbers.

However, before the piece was marked with the broad arrow, it had to be proved and examined.

Ordnance was subjected to several proofs before being accepted into service. Firstly, the piece was gauged internally and externally at

31 *The monogram of Henry, Earl Mulgrave, Master-General of the Ordnance 1810–18, as found on the barrels of ordnance*

49

several given points and the positions of the bore, chamber, vent and trunnions were checked for accuracy. The next test involved the firing of the pieces with pre-measured charges and shot, following which they were 'searched' or thoroughly inspected for any cracks or flaws in the casting. Table 8 gives a few examples of the charges used in various guns.

Table 8

PROVING CHARGES

Brass	Charge
Heavy 12 pdr	5 lb
Medium 12 pdr	4 lb
9 pdr	3 lb
Long 6 pdr	3 lb
Light 6 pdr	2 lb
Long 3 pdr	1½ lb
Light 3 pdr	1 lb

Iron	Charge
32 pdr	21½ lb
24 pdr	18 lb
18 pdr	15 lb

Brass guns were fired twice except light 6 and 3 pdrs which were fired three times. In iron guns, one shot and three wads were used and two rounds were fired per gun.

The third test to which ordnance was subjected was a water test. The barrel was chained in position with a pipe leading from a powerful water pump firmly fixed into the muzzle. The barrel was thoroughly cleaned on the outside and the vent blocked. Once the pump started, the examining officials kept a careful eye on the barrel to see if any water was forced through any minute cracks from the bore to the outside. On their visit in 1799 to the fleet at Plymouth, Sir William Congreve and Sir Thomas Blomefield inspected the guns

32 *East India Company ordnance, early nineteenth century. Left to right: 4 2/5in howitzer, 24pdr gun, 8in howitzer; all on travelling carriages*

and powder of the ships. On testing the guns, 496 were condemned.

Having passed these tests, the barrel was then marked with the broad arrow. It was not until 1857 that guns passing the ordnance tests were marked with the date, registry number and broad arrow on the first reinforce. Prior to this date there were no registry numbers and, as has been stated, the broad arrow was placed in a variety of places.

Carriages

There were three main types of carriage, depending on the size of barrel they mounted and the role they were called on to perform: garrison carriages, also used at sea, siege carriages and field carriages. There were also carronade carriages and mortar beds. In forts and fortified places the normal carriage was

33 12pdr gun and limber of Swedish design, c1790, showing the method of limbering and the type of limber. This style and method were in use throughout Europe by all countries

mounted on a traversing mount. Garrison and siege carriages were strongly constructed because of the weight of the piece and the recoil; because mobility was not important, they tended to be heavy. In field carriages, where mobility with lightness and strength were essential, more thought was given to the design.

The double bracket trail was used on all artillery, whether field or siege, before the introduction in 1792 of the block or single trail, designed by Lieutenant-Colonel Sir William Congreve. The first guns to be fitted with the block trail were 3 pdrs and the Royal

51

34 *9pdr brass gun mounted on a block trail carriage*

35 *Elevation of a brass 8in howitzer on a travelling carriage*

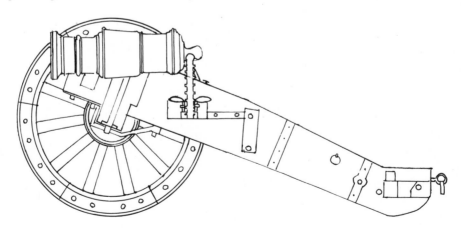

Horse Artillery adopted the block trail on its formation in 1793. The block trail was used on the 6 and 9 pdrs and was not adopted for the 12 pdr until much later.

The Royal Horse Artillery was equipped with two 12 pdrs, two 6 pdrs, and two $5\frac{1}{2}$ inch howitzers per troop. By 1800 the 12 pdr was no longer used, and so successful were the 9 pdrs in the Peninsula campaign that Sir Arthur Wellesley (later Duke of Wellington) ordered that more than half his Royal Horse Artillery should be armed with five 9 pdrs and a $5\frac{1}{2}$ inch howitzer per troop.

The block trail was not used by other artillery formations, which retained the double bracket trail. (This was still in use with some of the heavier pieces during the Crimean War and even the Indian Mutiny of 1857–8.)

A block trail was designed for the $5\frac{1}{2}$ inch howitzer, but was never manufactured. The 12 and 24 pdr howitzers, introduced in 1822, had a specially designed block trail carriage.

The most important considerations in carriages for field service were lightness, simplicity and conformity to pattern. Gauges and templates were made so that all parts, in theory, should be interchangeable. The wood was always well seasoned and the iron fittings made up from several pieces welded together, which gave added strength. The trail was made of oak, the cheeks of elm, the axle-tree bed of ash, the stocks of the wheels of elm, the spokes of oak. The fellies were ash for limbers as were the splinter bar and futchells, with the axle-tree bed of elm. The brackets behind were ash,

36 Brass gun mounted on a double bracket travelling carriage

37 9pdr brass gun mounted on a block trail with limber

the platform board oak, with the footboard of fir or elm and the prop stick of ash. Other woods were sometimes used, but in all cases the wood was selected for the task it had to do.

In four-wheeled carriages, the weight was equally distributed on each of the wheels, but where shafts were used in place of a pole the weight was on the back wheels. Block trail carriages travelled better and were more manageable when going downhill. The construction of the carriages can be seen from the illustrations.

Wheels for both gun and limber and carts were divided into classes, with, in each class, the piece, the limbers and wheels they were suitable for.

Limbers
1st class: 12 pdr gun and waggon.
2nd class: 9 pdr, long 6 and light 6 pdr, long 3

pdr heavy and light 5½ in howitzer and (after 1822) 12 and 24 pdr howitzers. Also for the ammunition waggon, forge waggon and wheel carriage.
3rd class: light 3 pdr.
The mountain service carriages had shafts only.
4th class: small arms ammunition waggon.
Wheels
1st class: 12 pdr gun and limber; 9 pdr, 6 pdr, all howitzers.
2nd class: 9 pdr and long 6 pdr limbers, light 6 pdr gun and limber, howitzer limbers, and light and heavy 5½ in howitzer.
3rd class: light 3 pdr to 4⅖ howitzer had the 4 ft 4 in wheel.
4th class: mountain service 3 pdr, 4⅖ in howitzer to 1 pdr had 3 ft wheels.
Ammunition waggons had the same wheels as the limbers of the guns they accompanied.

Axle-trees varied in construction with each type of carriage, so that the wheel of the long 6 pdr would not fit on the axle of the light 6 pdr. Dimension of the various classes of wheels are given in Appendix 4.

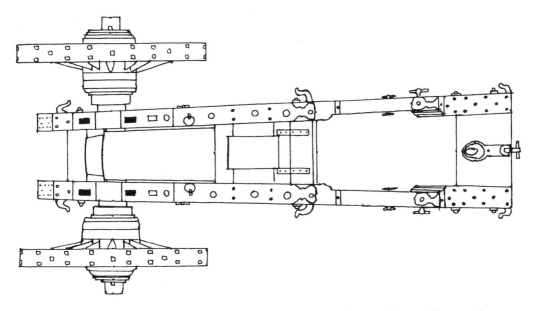

38 *Plan of a 24pdr travelling carriage*

39 *Elevation of an ammunition waggon*

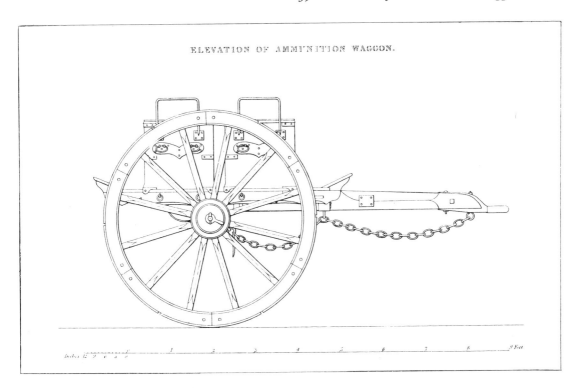

ELEVATION OF AMMUNITION WAGGON.

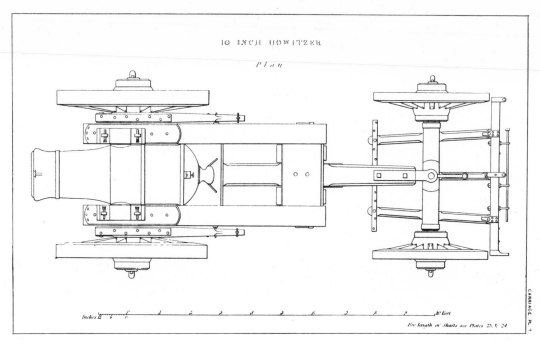

Plan

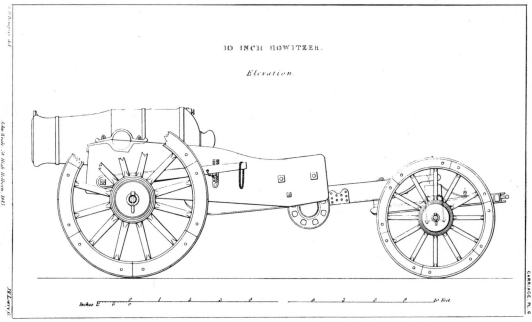

10 INCH HOWITZER.

Elevation.

40 *Plan and elevation of a 10in iron siege howitzer on a travelling carriage. From* Aide Memoire to the Military Sciences (*1845*). *Although the drawing is twenty five years later than 1820, the design had not altered*

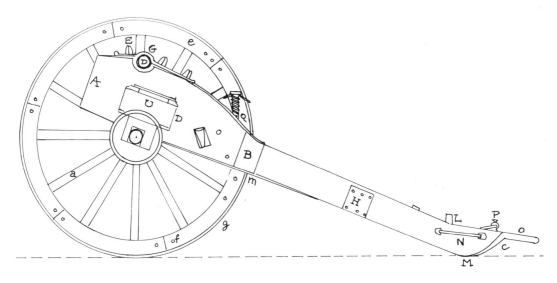

41 The names of the several parts of a block trail field carriage. AB, brackets; O, trail plate eye; BC, trail; D, trunnion hole; E, eye bolts, key and chain; G, capsquares; H, lock plate; L, traversing stay; M, trail plate; N, limbering-up handle; P, traversing loop; Q, elevating screw; U, match or shot box; a, spokes; e, fellies; f, felly rivets; g, tire or streaks; m, block

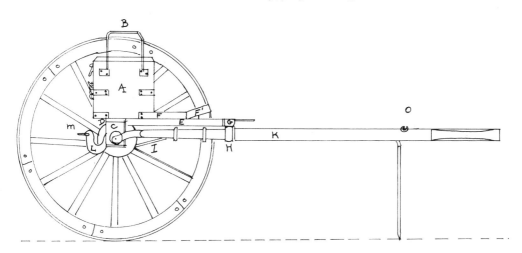

42 The names of the several parts of a limber. A, ammunition box; B, guard iron; C, axle-tree bed; D, lip of bed; E, futchell; F, foot boards; G, splinter bar; H, splinter bar socket; F1, platform dash; I, staff; K, shaft; L, limber hook; m, key; o, breeching loop

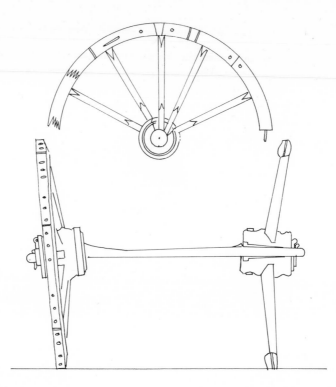

43   *The construction of a wheel*

The nave of the wheel was made of elm and was turned and bored before being hooped and having the mortises cut in for the spokes. The spokes were driven into the nave up to the shoulder but they were not set in perpendicular but standing away from it. This was called the dish, which was regulated by the size of wheel. Normally half an inch was allowed for every foot in diameter of the wheel. The wheel at this stage with just the spokes fitted was called a speech.

The fellies, of which there were six, were made from ash and were mortised with a dowling pin driven half-way into each at one end, leaving a hole for the dowl of the adjacent felly at the other. This was done to key the fellies together. The fellies were then fitted onto the speech, ensuring that the dowling pin on one felly entered the hole in the next. Once all the fellies were on, the spokes were split

at the ends and small wedges driven in to hold them tightly to the fellies. In this state the wheel was said to be 'on the wedge', and was left for some time to season. To prevent the dowling pins splitting the fellies, iron pins were driven in at right angles to the pins, close to each end.

The tyre of the wheel was constructed in six parts, called streaks, so made that any one could be removed and replaced in case of any damage. The streaks were put on white-hot and contracted as they cooled, holding the fellies close together.

Each streak had four bolts which were driven through the felly and secured by a nut on the other side. Two bolts were positioned, one at each end of the streak, and the other two close to the bolt that secured the dowling pin. Besides these, there were two large nails driven in near the spoke.

The axle-tree box, which was usually made of gunmetal, was hollowed out about an eighth of an inch in the centre so that only about three inches of the axle-tree rested on it. This prevented friction and served as a place to pack the grease. So that the spokes acted in a perpendicular direction, the axle-tree was bent downwards slightly. This was called the let.

Dishing added strength to the wheel, particularly when travelling over uneven ground when one wheel bore more of the total weight than the other. If there had been no dishing, the spokes and nave would probably have been forced out.

For garrison carriages and sea service carriages the wheels were called trucks. Trucks for garrison carriages were made of iron, while those of the sea service were made of wood. This was to avoid any damage to the decks as well as to facilitate any repair, which could be done by the ship's carpenter without carrying a stock of spare trucks. The wooden trucks were made from two semi-circular pieces of wood held together with an iron bar. Through the centre was a hole to take the axle. The trucks were then held in place with an iron or wooden peg which passed through a slot in the axle. All wheels less then twenty inches in diameter were termed trucks.

Mortar beds and carronade carriages were also strongly made, as mobility was not important in these instances.

44 A lower deck of HMS Victory *showing the various guns on wooden garrison carriages with wooden trucks. Also the tackle, rammers, buckets and shot*

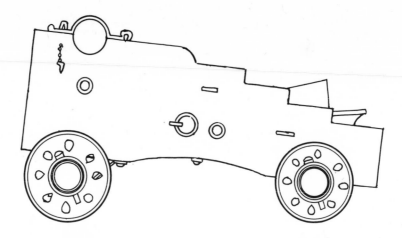

45 Wood garrison carriage with iron trucks, land service

In the sea service the carriages hardly varied at all, except for the mortars and carronades. They had two main defects, which were the awkward method of elevating the barrel and the difficulty in traversing the carriage, especially if the gun was of a large calibre.

Experiments were made with elevating screws but these never came into general use in the Navy during this period. Captain Robert Lawson reported on these newly designed carriages:

> A new garrison carriage with an elevating instrument has lately been experimented with which not only remedied these defects but renders platforms almost needless. Perhaps this construction might be useful for the lower decks of first and second rates, whose batteries, from their extent and steadiness, so much resemble those on shore.

In garrisons and forts after 1800, and during sieges, ordnance was fired from platforms. This prevented the piece sinking into the ground and provided a flat surface to run back on. The traversing platform consisted of a garrison carriage resting on two rails of wood, inclined at a slight angle. These rails were joined by bars and pivoted from the front in an arc, making traversing on a target quicker and less effort. In a lot of cases the rear trucks were removed from the carriages to reduce recoil. The rear trucks were replaced with chocks and the carriage became known as a rear chock carriage.

On board ship, the French guns had rear chock carriages from quite early on. They had no steps on them for elevating the barrel and they were said to have less recoil than a four-truck carriage. They were also said to be easier to traverse.

A large number of iron carriages were used in garrisons and forts, but these were not considered to be practical for use on board ship because of the extra weight and the fact that the cast iron would be liable to splinter if hit by an enemy shot. They were also not easy to repair and any major breakage put them out of action. Even in garrisons and forts they were used only in time of peace, a wooden carriage being supplied with every iron one for use in time of war.

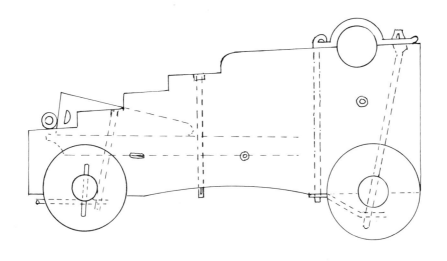

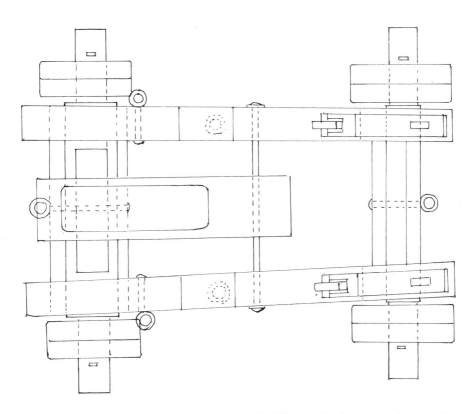

46 *Plan and elevation of a wooden garrison carriage with wood trucks for a 32pdr, sea service*

47 24pdr gun, cast in 1748, mounted on an iron garrison carriage with trucks

48 Naval gun with breeching loop, mounted on an iron garrison carriage with iron wheels

6 GUN EXERCISES

On land

A brigade of field artillery for active service in 1800 consisted of a company of artillery, a detachment of drivers and horses, six guns with an ammunition waggon for each one, a wheel carriage, a forge carriage, two store waggons and two spare ammunition waggons. In all there was a total of eighteen carriages including the guns. One gun with two carriages complete with its detachment and proportion of non-commissioned officers, drivers and horses was a subdivision, two of which formed a division and three, a half brigade. In the Royal Horse Artillery the unit was a troop which consisted of five guns and a howitzer together with the ammunition waggons etc. The troop could be divided into three divisions, two of two guns and one of one gun and the howitzer. These divisions could be divided again into sub-divisions each of which had one piece.

Nine pdrs and heavy howitzers had eight horses, and 6 pdrs and light howitzers had six horses. The other carriages had four horses. Drivers of the gun carriages, artificers, servants, grooms etc. were told off to their respective sub-divisions, every non-commissioned officer and man being provided with a blanket, haversack and canteen. The drivers had issued to them a forage cord and corn sacks or bags. Two tents and two camp kettles for each sub-division were carried on the waggons. The limber of the wheel carriage was usually filled with the wheeler's tools, the forge cart with horse medicines, and one of the waggon limbers was for the collar-maker.

The method of carrying the entrenching tools, picketing rope, tents, horseshoes, and other stores was regulated by the commanding officer according to the circumstances.

For drill purposes and exercise at home, four horses were the complement for each carriage of a six gun brigade, which consisted of one captain, three subalterns and the gun detachments with a sergeant-major of drivers, a trumpeter, three non-commissioned officers as markers, and two drivers to each of the guns and waggons. The senior subaltern was positioned on the right, the next in seniority on the left and the junior on the right of the centre division, dressing with the leading horses of the guns. The sergeant-major positioned himself on the right of the waggons covered by the senior corporal, and the sergeant of the drivers on the left of the waggons with the junior corporal on the right of number 3 waggon. The sub-divisions were then numbered off from the right. Numbers 1, 2 and 3 formed the right half brigade, numbers 4, 5 and 6 the left half brigade. Number 1 and 2 formed the right division, 3 and 4, the centre division, and 5 and 6 the left division. Guns 1, 3 and 5 were the right guns of the division and 2, 4 and 6 the left guns of the division. In all movements, number 3 gun was the gun of direction (the gun from which the others took direction or aim).

Plate 49 shows the type of horse harness in use from about 1800. The various component parts are detailed in the caption, these differing for the lead horses and those between or along the shaft of the limber. The near side lead horses in front of the shaft horse had saddles for the drivers. The leather was brown and up until 1853 each horse wore blinkers. This is not shown in the plate, which dates from 1859, but in general it shows the design in use during the period. The harness complete can be seen in plate 6, a team coming into action during the Peninsula campaign.

Distances between the guns and waggons were laid down depending on whether the guns were in line, on a flank, or in action, as follows.

Between guns and waggons in line
Fifteen yards between muzzle and muzzle and eighteen yards for horses, the waggons one horse's length in the rear and the same in the front when the line was retiring.

On a flank
Waggon horses one horse's length from the leader of the guns.

In action
Limber fifteen yards in rear of the gun, dressing on number 15 covered by the waggon.

All general words of command were immediately repeated by the officers commanding divisions. When line was ordered for action, the guns were loaded and fired by order of the commanding officer, but when the guns formed in succession, they loaded and fired as soon as they formed.

The loading and firing of the guns was a laid-down drill, as would be forming line in the infantry or any other military manoeuvre. Each man of a gun detachment was numbered off, with the non-commissioned officer normally as the number 1. The size of the detachment depended on the calibre of the piece, one man

being allowed for each 500 lb of metal.

The number 1 of the gun stood behind the trail with the ventsman and the firer on either side of the breech, and the loader and the spongeman positioned themselves in front of the axle of the gun, one on each side forward of the wheels.

When reloading after a shot had been fired, the spongeman wetted the sponge on the rammer in the bucket of water and swabbed out the bore. This was done to extinguish any small smouldering particles left from the previous cartridge, which might ignite the new charge prematurely. The loader then placed the new cartridge in the muzzle, and the spongeman with reversed rammer pushed it home. At the same time as this operation was being carried out the ventsman put his thumb over the vent to prevent the rush of air from the newly rammed charge rekindling any spark that might still be in the bore. This action was termed 'serving the vent'. Failure to do this on the part of the ventsman resulted in a sharp hit on the head from the spongeman, whose very life depended on the ventsman's thumb. A leather thumbstall was provided for the ventsman.

Next the projectile was loaded, although with compounded ammunition such as case shot the projectile and its charge were put in together. It was not uncommon, though, for the charge and projectile to be put in the muzzle, one after the other, and rammed home together. This speeded up the rate of fire. Once the charge and projectile had been rammed home, the ventsman pricked the bag of the charge by pushing the 'pricker' down the vent. The piece was then primed by the ventsman, the method depending on whether a tube, loose powder or cannon lock was in use. In the case of a tube or loose powder, the ventsman fired the primer with a linstock or a portfire; with a cannon lock, he

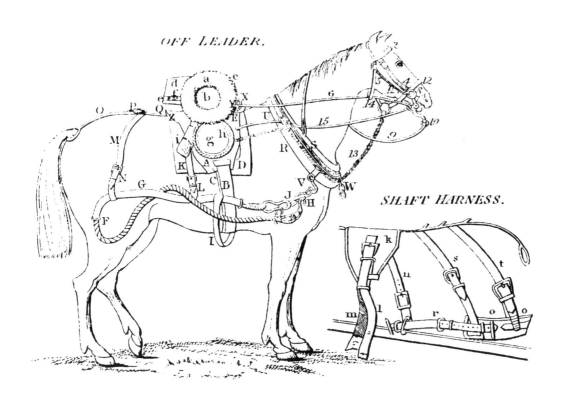

OFF LEADER.

SHAFT HARNESS.

49 Type of harness in use, from a book published 1859; the same design in brown leather was used during 1790–1820 but with the addition of blinkers

OFF LEADER

A, pad or off saddle; B, surcingle; C, girth of the pad; D, pannel of the pad; E, pad staples; F, trace; G, pipe of the trace; H, hook of the trace; I, belly band of the trace; J, trace links; K, bearing strap; L, buckling piece of bearing strap; M, hip strap; N, buckling piece of hip strap; O, crupper; P, crupper ring; Q, buckling piece of crupper; R, collar; S, the hames; U, the housing strap; V, shoulder link and hook;

W, breast chain or link; X, wither strap; Y, bearing hook; Z, cantle of pad; a, sheepskin; b, valise; c, baggage strap; d, mess tin; e, mess tin strap; f, buckle of mess tin strap; g, wooden canteen; h, forage cord; i, nose bag

BRIDLE

2, front or brow band; 3, cheek; 4, cheek billets; 5, throat lash; 6, bearing rein; 7, bit; 8, cheek of bit; 9, leading rein; 10, bar of the bit; 11, head collar; 12, nose band; 13, collar chain; 14, jowl strap; 15, side rein

SHAFT HARNESS

k, pad or off saddle; l, back band; m, shaft tugs; n, bearing strap; o, breeching; r, strap of breeching; s, loin strap; t, hips strap

50   *Loading drill for a field gun*

pulled on the lanyard to activate the mechanism. Depending on the size of the gun, extra gunners handled the charges and projectiles and passed them to the loader. In action, when a linstock was used, it was normally lit on the word of command to prepare for action, and would be kept burning throughout the action.

There were also other drills laid down for heavy artillery, garrison artillery and mortars. In all garrison drill, the men not actually employed on the gun fell in under cover of the parapet, their backs to the wall and in line with the edge of the gun platform, the odd numbers on the left and the even numbers on the right. When running up and back the gun (running up to fire, running back to load), the men were never allowed to position themselves behind the handspikes because an enemy shot could hit it and probably kill or wound a man. Garrison guns could be served by any number of men from three to eight. With less men, the gunners doubled up on the jobs that had to be done. In ramming the sponger kept his body well back in case of any premature explosion, which would then only affect the hands. Every three or four hours a

wad hook was used to clear away any pieces of cartridge bags left in the bore.

The drill for a gun on a travelling carriage was the same as garrison drill except that in running up the gun, numbers 3 and 4 placed their handspikes over one of the wheel spokes and under the bracket of the carriage. In traversing the gun, number 8 placed a crooked handspike at the end of the trail so that the gun could be moved left or right, depending on where the target was.

When the garrison gun was on a traversing platform, numbers 1 and 2 placed themselves in front, number 3 handing the sponge and wad hook to number 1 while number 4 supplied number 2 with ammunition. Number 5 served the vent, standing on the platform in front of the rear truck. Numbers 2, 4, 5 and 6 were responsible for running out the gun. If the pivot of the traversing platform was at the front the duties of numbers 5 and 6 were traversing and those of numbers 3 and 4, elevating. With the pivot at the rear, these numbers reversed their duties. When the gun

51 Garrison carriage mounted on a traversing platform

*52　A devil's carriage, showing position of
garrison carriage and barrel*

*53　Platform carriage, showing position of
barrel and garrison carriage*

was ready to fire, number 6 stood on the ground if a lock was used, or on the platform if a portfire was used.

The drill with mortars was somewhat different to that for garrison guns. During the loading numbers 1 and 2 stood on the mortar bed and helped numbers 3 and 4 putting in the shells by placing their arms under the supporting bar. So as to find out what angle the mortar was being elevated to, a plumb line was held over the end of the trunnions where the different degrees of elevation were marked. The mortar was then laid on the target by a man standing behind the piece with a plumb-line, the others traversing the bed until the barrel was pointed at the target. The line of metal on the mortar was made by marking two points, one on the base and the other on the muzzle ring, and connecting the two with a line of chalk.

The method of 'working' carronades in the land service employed five men. Numbers 1 and 3 positioned themselves on the right and numbers 2 and 4 on the left. Number 1 sponged and rammed home the charge and projectile, while number 2 loaded and number 3 served the vent which was fired by number 4. In running up, numbers 1 and 2 fastened the hooks at the end of the double blocks to the rings on the parapet. When traversing, numbers 3 and 4 fastened ropes to a pin on each side of the carriage and pulled on the command 'traverse' to either the left or the right.

There were also many laid-down drills for mounting and dismounting ordnance using gyns, and moving ordnance from travelling carriages to their firing position. When using gyns the number of men was usually about eleven, but without a gyn about thirty men were required to mount and dismount ordnance from their carriages.

The transport of heavy pieces not on travel-

54   Two-wheeled sling cart with barrel slung in position

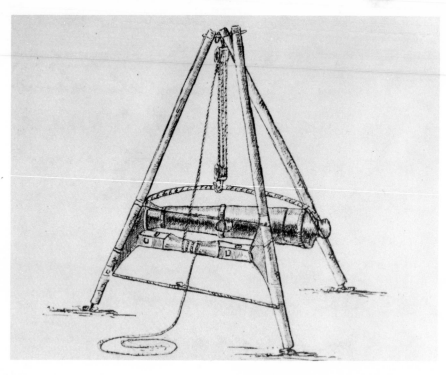

55   *A triangle gyn*

ling carriages, such as mortars and garrison guns on garrison carriages, was accomplished by the use of a devil's carriage, sling cart or platform carriage.

To mount a garrison gun and its carriage on a devil's carriage needed sixteen men. Numbers 1 to 8 were drag-rope men, 9 to 14 were perchmen and 15 and 16 were preventative men. At the command 'limber up', drag-rope men and preventative men pulled down the perch, 13 hooked on the draught chain and 14 hooked up the limbering chain. Before this could be done, the gun and carriage were turned over and the carriage raised on its front until clear of the trunnions. The devil's carriage was run forward until the rear axletree was over the trunnions. The carriage was then lashed to the perch and the gun slung underneath with a chain in front of one trunnion and behind the other. The perch was

pulled down on the limber and the gun depressed by pushing down on the handspikes which had been put in the bore of the barrel. The breech was then lashed to the perch. Mortars were also slung under a devil's carriage, the 13 inch mortar requiring two carriages.

The sling cart was used for carrying heavy ordnance short distances. The barrel of the gun was slung beneath the axle of the cart and lashed tight. When slinging mortars, the beds were slung on a separate carriage, but with smaller mortars the barrel and bed were slung from the same carriage.

Besides gyns, there were other drills for making cranes from carriages and limbers of guns for embarking stores and using wheels to raise a gun on a sling cart up a hill. These exercises were intended for active service when the proper equipment was not always available.

In the Royal Artillery great care was taken

Table 9

DUTIES IN GARRISON GUN DRILL

No. of men in detachment:	8	7	6	5	4	3
Searches and sponges	1	1	1	1	1	1
Loads and helps to run up	2	2	2	2	2	2
Runs up	3, 4, 5, 6	3, 4, 5, 6	1, 2, 3, 4	1, 2, 3, 4	2, 3, 4	1, 2, 3
Elevates	3, 4	3, 4	1, 2	1, 2	2, 3	1, 2
Traverses	5, 6	5, 6	3, 4	3, 4	2, 3	1, 2
Serves and primes the vent	6	6	4	4	4	3
Fires	7	4	5	2	2	2
Brings cartridge	7	4	5	2	2	2
Points and commands	8	7	6	5	4	3

Table 10

DRILL WITH MORTARS

	10 or 13	10 or 13	8 or 5½
Number of men	7	12	5
Sponges, uncaps the fuse, makes the shell, and helps to ram in	1	1	1
Loads and helps put shell in	2	2	2
Run up	1, 2, 3, 4, 5	1, 2, 3, 4, 5, 6	1, 2, 3, 4
Brings cartridge	2	8	2
Brings shell	3, 4	3, 4	3
Traverse	3, 4, 5, 6	3, 4, 5, 6	3, 4
Serves the vent	5	5	4
Primes	5	5	3
Fires	6	6	4
Points	7	7	5

to ensure that all guns had their full complement of equipment and that each man was so trained as to be able to take over another man's duties should he be killed or wounded.

Drill duties with gun crews of varying numbers for garrison guns and mortars are shown in Tables 9 and 10.

At sea

At sea the gun drill was far simpler than on land, as the only guns were mounted on garrison carriages, except for the carronades and the mortars on board bomb vessels. Each calibre of ordnance, as on land, had a certain number of men to handle it. In the Royal Navy there were the captain of the gun, a certain number of seamen, and a seaman or boy to fetch the ammunition. The number of seamen varied for each calibre as follows:

36pdr	24pdr	18pdr	12pdr	8pdr	6pdr	4pdr
12	10	8	8	6	4	4

The captain of the gun was responsible for the pointing of the piece and placed himself behind the breech. The seamen divided themselves equally on each side of the gun with the first man on the right sponging and ramming home the charge. The first man on the left was the loader who took the charges from the powder-man whose job it was to fetch these during action. In the main, these powder-men were boys. The last man on the right-hand side of the gun wore a small apron with a pocket holding some spare flints for the cannon lock and an oily rag for cleaning it.

On the larger ships of the line, two gunner's mates in each battery were provided with tool bags containing spare locks, screw-drivers, spare lanyards for the gun locks, etc. On the smaller ships of less than ten guns, one of the gunners carried the tool bag.

Ships' guns were in the charge of lieutenants and midshipmen. On principal gun-decks, two lieutenants would share the responsibility and each group of, say, half-a-dozen guns would be under the eye of a midshipman. There would, of course, be many occasions when there were insufficient junior officers, but the problem of control was of the utmost importance and captains would, no doubt, ensure that the regulating of fire was maintained as fully as possible. As guns did not fire simultaneously, partly because only some would bear and also because of damage to the ship which would result from firing too many guns at once, controlling officers would have time to pass from group to group to ensure that guns fired at the most opportune time.

The Gunner, a warrant officer well known for his 'steadiness', worked in the magazine and supervised the issue of ammunition. He was assisted by Mates who were petty officers of some considerable experience. The Gunner, though a warrant officer, was junior to the Purser, Surgeon and Master. Besides his duties as Gunner, he was responsible for the younger 'young gentlemen', King's Letter Boys, the Captain's and other servants and junior midshipmen. They all lived in the Gun Room under the Gunner's charge until about the age of 15, when they were moved to the after cockpit to join the senior midshipmen and master's mates.

The signal to clear the ship for action was given by the 'beating to quarters' on the drum.

As soon as the drumrolls started, the powder-men went to the Gunner's store to get the priming horns, tubes and cartridge boxes. The last two men on the right side of the gun fetched the locks and any other articles needed for the guns. The rest of the men got the guns ready for action by casting them loose and removing any obstacles on the deck. The ropes were then hooked up to the carriage. The guns did not normally require loading as it was the practice to load immediately the ship left harbour. Every member of the gun crew also had other tasks such as boarding-party, firemen, sail-trimmers, and men who raised and lowered the gun ports.

On a large ship such as the *Victory*, the gunpowder was kept in barrels and ready-made charges in three magazines. There was no light in the magazine, light being provided from an outside lamp that shone through a window. In the magazine, the men who worked there wore felt slippers and used only copper tools for opening gunpowder barrels. These were normally made of wood, but in 1810 James Walker patented a 'vessel for the safe conveyance' of gunpowder which was made of copper or brass and strengthened with hoops. It appears that Sir William Congreve was directed to report on the usefulness of these barrels, and in August 1811 he reported that they were not suitable for the Royal Navy. Congreve then began to make barrels lined with copper or lead and to supply them to the Navy. This infringed Walker's patent and he successfully brought an action against the government for loss of profit, which it appears he won in 1816.

The gun deck was served from the grand magazine, the middle deck from the fore magazine, and the main deck, quarter deck and forecastle from the aft magazine. There were strict rules that new cartridges could be given only to powder-men bringing back the empty cases. On the decks were positioned match tubs which were half-full of water and had slow matches in the notches round the rim burning without any risk of igniting the powder.

The first thing that had to be done was to remove the tampions, which were muzzle stops, before firing could take place. In order to explain more easily the movements involved in the loading and firing of the guns, suppose that the first shot had been fired. The recoil of the gun had thrust it back against the breeching rope, which passed through a ring on the cascable of the barrel. The thick rope was secured to a ring on one side of the gun port, and was passed through a ring on the carriage, another on the cascable, and another on the opposite side of the carriage, before being secured to another ring on the opposite side of the gun port. This acted as a rudimentary form of recoil brake. The cartridge was then got ready while the captain of the gun put the priming wire into the vent of the gun to see if it was clear. The first man on the right side took the sponge from the man next to him and swabbed out the bore. Withdrawing the sponge, he struck it several times on the muzzle of the gun to remove any particles of burning or smouldering cartridge case from the last round. The captain of the gun then 'served the vent' while the last man on the right side cleaned the lock and half-cocked it.

The cartridge was then put in the muzzle and rammed home. The powder-man, once he had handed the charge over, immediately returned to the magazine to fetch another, as it was dangerous to have too many cartridges stacked by the guns during action. A wad followed the cartridge and then the projectile followed by another wad. The captain pushed the pricker into the vent to pierce the bag of the cartridge, and then gave the order to run out the gun.

The rear man on the left loosened the train tackle while the rest of the crew took up on the side tackle. On word 'heave', all the side tackle men pulled and the gun ran out. The gun was then pointed by the use of the side tackles and handspikes under the carriage. The next operation was the elevating of the gun, and for this two men raised the barrel with a crow-bar resting on the steps of the carriage while a third man pushed the quoin in further. The train tackle had by this time been disconnected from the ring in the deck, and the side tackle men held the ropes, waiting for the gun to fire, when they would drop the ropes. The captain positioned himself a suitable distance behind the gun so as to be out of the way of the recoil and on the command to fire pulled the lanyard of the lock. If the gun lock broke or failed to work, the firing was carried out by using quill tubes, as long as there were any available, otherwise priming powder was poured on the vent, and a portfire used.

Ships in battle were usually engaged on one side only, each gun manned with a full crew on either port or starboard. If, during the engagement, the ship was required to fire on both sides, the crews of guns number 1, 3, 5, 7, etc. stayed on the starboard side, for example, and the even numbers moved to the port side. Each crew would then work its own and the adjacent piece. If all crews were suddenly reordered to one side, the crews that were to move loaded the guns and ran them out, except on the lower deck where they were prevented from running out by securing the tackle and putting a crow bar under the trucks of the carriage.

Because of either boarding-parties, fire on board ship, or a need for sailors to increase the musketry fire or to handle the ship because of casualties, half the gun crews' number might be taken to fill their allotted secondary task in action. Under these circumstances, the crews of two adjoining guns would combine and work the pieces alternately. If more men were taken, then three or four guns would combine, again firing in succession.

The loading and firing of carronades was not very different from the procedure for the larger guns. A quoin was always kept by the carronade in case the screw elevating gear was either damaged or jammed, or it was necessary to elevate the piece higher than was possible with the elevating screw. The crew of a carronade consisted of a captain, two seamen and a powder-man. Twelve pdrs could be handled by only two men, 18 and 24 pdrs by three men, and the 68 pdr required four men. Just before the firing of the piece, the man on the left always took up the portfire and held it ready over the vent should the lock, when fired, fail to spark.

When the sea was so rough that the lower deck ports had to be closed, or when the ship was close to that of the enemy, the wooden-shafted sponge and rammer was dispensed with and in its place the rammer and sponge with a thick rope shaft was used. This was easier to handle in the confined space below decks.

Guns were hardly ever loaded with more than one projectile, and carronades never with more than one. If, by special command, guns were required to fire more than one projectile at once, then this double load usually consisted of two round shot or one bar shot and one round, or one grape and one round. The round shot was always loaded last because it had greater velocity than either bar or grape.

7 BOMB KETCHES AND FIRESHIPS

Bomb ketches

Bomb vessels, or 'bombs' as they were sometimes called, first appear to have been used by the French, when they constructed special ships to take mortars. It was discovered that a ship with conventional armament could not get close enough to a fort on shore for its guns to have any effect and that the fort's guns were capable of pulverising any ship audacious enough to come in too close. Mortars could throw a shot a considerable distance, but they could not be housed on a normal ship because the elevation at which they fired would send the shot into the sails and rigging of the ship.

Earlier bomb vessels had a main mast and mizzen mast but no foremast: the idea was to lie the ship bow-on to its target and fire the mortar over the bows. In the early nineteenth century, bombs were constructed with three masts and lay broadside on to the target. Although this presented more of the vessel for the opposing gunners to fire at, fortress guns did not possess the range. The use of three masts made the vessel more manoeuvrable.

The British used four bomb vessels in the attack on St-Malo in 1693, and in 1703 the Ordnance was commanded to equip a bomb vessel with two 13 inch mortars. Bomb ships were very strongly built, because of the weight of the mortar and the force of the recoil, and had a shallow draft for inshore work. Each

ship normally carried two mortars, one of 13 inches and the other of 10 inches. These mortars were considerably larger than their land counterparts of the same calibres, as they were required to range further. At the end of the eighteenth century, four 68 pdr carronades and six 18 pdr carronades replaced the 13 inch mortar on board bombs, but the larger carronades seem to have had a short life on the bomb vessels as they were themselves soon replaced by the mortar they had ousted.

The bed or carriage of the sea service mortar was completely different from those used on land. On board the bombs, the bed had a circular base, on which the mortar traversed, and it was fixed to the main deck of the ship by large iron bolts, called pintles, which passed through the deck and keyed on a frame or timber supporting the deck. Beneath each mortar bed was the shell room for that mortar. The ship itself was constructed with three masts which were placed in a position so as not to be affected by the firing of the mortars.

Mortars fired at an elevation of 45 degrees, which was supposed to give the greatest range and accuracy. A detachment from the Royal Artillery, consisting of one subaltern and ten men together with a carpenter from the Tower of London, manned the guns. A smaller ship of about 100 tons was ordered by the Board of Ordnance to accompany each bomb

56 *A sectional model showing the interior of a bomb vessel with reinforced supports for the mortars*

vessel to carry some of the detachment and stores for the mortar.

The practice of using Royal Artillery and other Army personnel was discontinued in 1804 with the formation of the Royal Marine Artillery. Difficulties had been encountered in getting military and naval officers to work together and there was also some argument as to whether military officers should, when embarked on board a 'bomb', be subject to naval discipline. A typical example was quoted by Captain Brenton concerning the bombardment of Havre-le-Grace in August 1804:

> The bombardment is memorable only for the dispute which arose between the officers of the Navy and the Artillerymen embarked on board the Bomb vessels. The privates refused to do any other duty than simply that of attending to the mortars in time of action and keeping them prepared for service; their officers supported them in this determination, and the commanders of the Bombs appealed to the Admiralty in consequence of which the Marine Artillery was formed and embarked in the Bombs.

The Royal Marine Artillery was established from the Order in Council of 18 August 1804.

Table 11

DIMENSIONS OF MORTARS AND THEIR BEDS

Mortar	Length (ft in)	Weight (lb)	
13 in	5 3	82	
10 in	4 8	33	
Bed			
13 in	7 10		depth 2 ft 11 in, breadth 4 ft 6 in
10 in	7 0		depth 2 ft 5 in, breadth 3 ft 11 in

Table 12

EXPERIMENTS CARRIED OUT AT WOOLWICH JULY 1798 TO FIND THE RANGE OF SEA SERVICE MORTARS AT 45 DEGREES

13 inch mortar			10 inch mortar			
Charge (lb)	Range (yds)	Time (secs)	Charge (lb)	Range (yds)	Time (secs)	Fuse (in)
2	690	13	1	680	13	2·7
4	1,400	18	2	1,340	18	3·75
6	1,900	21	3	1,900	21	4·36
8	2,575	24¾	4	2,500	24½	5·09
10	2,975	26½	5	2,800	26	5·5
12	3,500	29	6	3,200	27	6·02
14	3,860	29½	7	3,500	29	6·13
16	3,900	30	8	3,800	30	6·23
18	4,000	30½	9	3,900	30¼	6·33
20	4,200	31	9½	4,000	30½	6·44

The ordnance stores for a bomb vessel were as follows:

Shells for 13 inch mortar	200
Shells for 10 inch mortar	200
Carcasses, round, for 13 inch mortar	140
Carcasses, oblong, for 10 inch mortar	40
Round shot of 1 lb	5,000
Barrels of gunpowder	240

There were two types of carcass for setting fire to buildings and ships. Those for the 13 inch mortar were cast as ordinary shells but had five vents instead of the normal one vent. The other type was oblong, resembling an egg, and was made from two spherical iron cups joined together by iron ribs and then bound with canvas and cordage. In the 10 inch carcass there were sometimes eight pistol barrels loaded with two balls each, put inside amongst the mixture. These fired in different directions during the time the carcass burned. The oblong carcasses did not have the range or the accuracy of the round ones.

The 5,000 1 lb shot were first added to the stores of bombs in 1758. They were always called pound shot and were fired from the mortar loose. When the mortar's chamber had been loaded with powder, a wooden plug was placed over the entrance to the chamber and a number of pound shot poured in loose on top. These shots, when fired, came down on the target in a shower from which it was said to be difficult to find cover. This form of shot proved to be very efficient against open works or lines where there were no casemates. A further use of this type of shot was to protect landing parties or re-embarking troops. The usual load

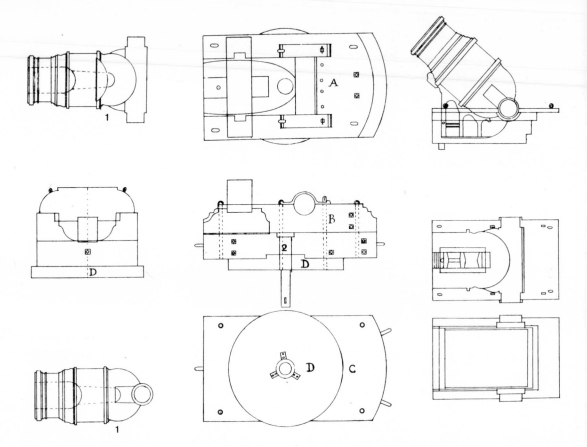

57   Names of the parts of a sea service mortar bed. A, upper side of bed or seat; B, section of the side; C, bottom of the bed; D, circular base; 1, the mortar; 2, the pintle

for a 13 inch mortar was 200, and for a 10 inch mortar, 100. Pound shot was not used over 1,500 yards, because the spread was too great to be effective. The use of the correct powder charge was particularly important when using compounded ammunition, as otherwise the effect would be minimal.

There were strict precautions taken on board the bomb vessels to minimise the risk of fire or explosion. No powder barrels were opened, nor were any charges measured out, except in the captain's cabin. The door was kept closed and covered with tanned hide.

The cabin and the passage leading from it were kept constantly watered. There was also a bulkhead of three-inch-thick wooden planks between mortar and magazine. Wet curtains where hung over the doors and hatchways during action and a wooden screen was placed behind the vent of the mortar when it was fired.

Bomb vessels were nearly always named after a volcano or a name connected with fire and explosions. Typical names were *Serpent, Grenade, Sulphur, Thunder* and *Strombolo.*

Fireships
The use of fireships precedes the use of any other form of artillery. The ancient Greeks are recorded as having used ships loaded with blazing faggots in their sea battles. Fireships were used by the Crusaders in the capture of

Constantinople in 1204. The discovery of gunpowder made these vessels more destructive than before, and they were used with great success against the Spanish Armada. In 1636 the Navy considered the fireship such an important weapon that they were made a permanent part of the fleet, with a crew of one master, five petty officers and twenty-five seamen. Until 1693 fireships had been used solely for setting fire to opposing ships, but in that year at the attack on St-Malo a ship armed like a floating bomb was first used. The designer was a Frenchman named Renaud, who was also credited with the invention of 'bomb' vessels. The *Infernal* was a 300 ton ship, 33 feet long with a 9 foot draft. It was ballasted with sand and crammed with explosive material. On the lower deck were 200 barrels of gunpowder, on the next deck 600 bombs and carcasses, on the next deck barrels of fireworks and grenades; the upper deck was covered with old iron cannons and scrap metal. The explosion was said to have shaken the whole town and broken down part of the wall.

Fireships were not normally loaded as heavily as the *Infernal* but were carefully designed so that the fire would spread to every part of the ship once ignited. Great care had to be taken with these ships, in their preparation as well as in their use. Wind and current had to be correct if they were to do their work properly. The fitting out of a fireship was both expensive and tedious, and its use was therefore confined to the rare occasions when it could do the most damage to an enemy fleet.

The vessel selected, usually a purchased merchantman, was prepared in the naval dockyard. Part of the lower deck, including the masts, pumps, and hatchways, was partitioned off and sealed by a forecastle bulkhead. Another bulkhead was constructed behind the main chains stretching the width of the ship. All the space outside this partition

was called the fire room, where the highly inflammable material was stored. In this room there were a number of troughs which ran fore and aft and across the ship. These were raised above the level of the deck and liberally covered with pitch and resin.

Six or eight small ports were cut in each side of the ship with their shutters falling down instead of opening upwards, which was the normal procedure. On each side there was also a large port with folding shutters which was termed the sally port. These were situated on the quarter deck of the ship and were connected by wooden troughs through the aft bulkhead to the fire room. In the upper deck, eight small scuttles were cut each side, and these again communicated with the fire room. Those that were under the shrouds had a wooden chimney which was called a fire trunk.

The stores for a fireship were prepared by the Royal Artillery, and as soon as the ship was ready an officer went on board to supervise the placing of the material. In each fireship there were the following stores:

Fire barrels	8
Reeds, long	158
Reeds, short	150
Bavins	200
Curtains	38
Composition for priming	$1\frac{1}{2}$ barrels
Quick match for priming	1 barrel
Portfires	24
Chambers for blowing open ports	12 or 16
Hand grenades	60

Each of the fire barrels contained about 400 lb of highly inflammable material made from Swedish pitch, tallow, saltpetre and corned gunpowder. The reeds, tied in bundles; the bavins, which were made from small birch twigs; and the curtains, pieces of canvas about a yard square, were soaked in a mixture of Swedish pitch, resin, sulphur and mealed gunpowder. The chambers were small iron sockets which were loaded with gunpowder and were used to blow open the ports.

The material was placed about the ship in the following way. The longest reeds were placed in the fore and aft wooden troughs, and the shorter ones in the troughs that ran across the width of the ship. The fire barrels, which were filled with inflammable material, suitably fused, were designed to carry the fire aloft. Four were placed each side under the fire trunks and the other four under the scuttles between them. The bavins were then tied over the reeds, and the curtains were nailed to the beams, above the troughs that ran across the ship. Some reeds were tied upright at the corners of the fire rooms and around the fire barrels.

The ports were next caulked up and an iron chamber fixed against each one. Everything was securely tied down so that the ship was in readiness.

When the order for firing the ship was given, the barrels containing the priming powder were opened and the contents sprinkled thickly in the troughs under the reeds. More powder was then put on top of the reeds and quick match. Lastly, the bavins were put loosely in all the troughs. The fire barrels were opened and the quick match which had been placed inside were drawn out and hung over the edge of the barrel. Fuses of quick match connected the troughs to the fire barrels and ran from the barrels to the freshly primed vents of the chambers. The two communication troughs between the fire room and the sally ports were sprinkled with powder and ten quick match fuses were laid. A strong fuse connected the sally port from which the ship was going to be fired to the other one so that the fire would burn equally on both sides of the ship.

The remaining quick match and primer were placed about the fire room so that 'no interruption may happen to the passage of the flames in any part of it'. The plugs and covers were taken off the scuttles and fire trunks so that the flames would be able to reach the upper parts of the ships and the rigging. Lastly, hand grenades were scattered about the deck to deter the enemy from any attempt to board the ship before it had been effective. Portfires were securely fixed to the fuses at the sally port, and the ship was ready.

As soon as the final signal was given, the ship was pointed at the target, the sails set and the helm fixed. The crew, except for the captain, then took to the small boats which carried them to the covering ships, while the captain lit the portfires and hurriedly left the ship by the sally port.

Great care had to be taken to ensure that there was no gunpowder left in barrels or confined spaces as this would result in the ship blowing up instead of burning. All the stores on board a fireship were calculated to burn and not explode. It was most important that the crew appreciated this danger and checked everything thoroughly when preparing for action.

Besides properly fitted out fireships there were many other smaller craft used for the same purpose. These were used in harbours and rivers and were normally made on the spot for a particular job. They were called fire stages or rafts and consisted of timbers piled with any inflammable material that could be found and covered with more wood. Boats loaded with gunpowder with the fuses wound round a cannon lock were also used. A string was tied to the firing mechanism of the gun lock which was cocked and primed. The boat floated down on the current or tide, with the string either held by someone in another boat some distance away or hanging loose so as to catch in bushes on the shore or in driftwood. The jerk of the string fired the cannon lock which fired the gunpowder, blowing up anything near the boat.

When operations of this type were undertaken during the day, boats or floats loaded with barrels of smoke composition were sometimes used to cover the approach of a fireship or small explosive boat.

Although fireships had enjoyed popularity during the early part of the eighteenth century (they had been mentioned in the Navy List since Charles II), only one such vessel was listed in the Navy List of 1777–8. However, by 1782 there were eighteen listed. After the Peace of 1783 between Britain, France, Holland, Spain and United States the number fell to four, rising to nine by 1790. The number of fireships fluctuated during the early part of the Napoleonic wars, reaching a peak of sixteen in 1805. By 1820, they had virtually disappeared, Britain's sea supremacy after Trafalgar making them obsolescent.

APPENDICES

During the period 1790–1820 there were a number of patents filed in relation to artillery and gunpowder. The largest number in any category was nine, which were either for carriages or mentioned new forms of carriages in the specification. As the majority of these inventions were never adopted for use, it is intended to list them with a brief description of the main points of the invention under various headings.

Carriages

8 December 1796. John Glover. No 2,151

Describes an under carriage with or without trucks with an adjustable incline, the top carriage sliding on the under one.

20 March 1798. Joseph Haycraft. No 2,224

Describes a carriage in two parts, the upper sliding on the inclined surface of the lower. The piece is traversed by a conical wheel at right angles to the rollers of the lower carriage.

24 July 1800. Anthony Cesari de Poggi. No 2,428

Deals not only with sliding carriages and a carriage with the cheeks in the form of a quadrant of a circle, but with apparatus for checking recoil, polygonal trunnion holes, screw apparatus for elevating, and a conical chambered mortar.

19 December 1805. John Glover. No 2,803

Describes improvements to the patent of 8 December 1796, above.

13 March 1806. Michael Logan. No 2,917

Deals with the construction of a gun carriage containing the minimum of material and presenting the smallest target to the enemy. The carriage could be made of iron or wood.

29 January 1807. Richard Friend. No 3,005

Describes a carriage made to slide on a normal carronade slide carriage.

8 April 1807. Richard Francis Hawkins. No 3,028

Guns were run out by ropes attached to barrels or drums and worked by a winch with a ratchet.

24 May 1808. William Congreve. No 3,134

Describes a simple gun carriage made by casting guns with four trunnions and placing a truck on each, or two trunnions with a truck on each and the cascable supported by a frame running on another truck. The trucks could be locked to lessen the recoil.

11 May 1812. William Congreve. No 3,565

Involves casting ordnance in the same way as carronades, i.e. with a loop instead of trunnions. The carriage was in two parts, an outer carriage with a traversing motion and an upper or inner carriage which recoiled and slid on the other. The trucks were round or polygonal and could be locked when fired so that the top carriage slid.

Gunpowder, cartridge and shot

13 November 1802. Henry Smith. No 2,658

Describes barrels for the safe carriage of gunpowder. The barrels were made of copper and were watertight. Larger fixed barrels had a tap by which any quantity of powder could be drawn.

12 November 1807. John Dickenson. No 3,080

Describes the manufacture of cannon cartridge paper. A certain amount of wool or woollen rags was to be mixed with the linen in making the paper, to prevent sparks being retained after the charge was fired.

11 April 1810. William Parr. No 3,328

An improvement relating to the manufacture of gunpowder. No specification enrolled.

7 September 1810. James Walker. No 3,373

An improved vessel for carrying gunpowder. The vessels were made of copper or brass and strengthened with hoops (see Chapter 6 above).

4 December 1811. Frederick Albert Windsor. No 3,510

The method for making gunpowder employed sugar in the mixture. The proportions were 25 parts of sugar to 100 parts of gunpowder.

3 July 1815. Sir William Congreve. No 3,937

Gunpowder, new method of manufacture.

Miscellaneous

2 August 1800. Thomas Gill. No 2,436

Describes a new method of rifling guns.

23 January 1801. Robert Vazie. No 2,466

For improvements in the construction and application of guns. Describes a furnace for working a forge hammer, a breech-loading cannon, and a saddle for mounting a gun on the back of a horse. Special clamped shoes to enable the horse to stand the shock of the recoil are also described.

23 March 1803. Durs Egg. No 2,692

For improvements to firearms and locks, but mentions that cannon may be fired by discharging a pistol at the touch hole.

30 July 1808. George Richards. No 3,155

Describes a method of making single and double cannon and carronades. Also describes a breech-loading method.

26 July 1811. Henry James and John Jones. No 3,469

An improvement in the manufacture of all types of barrels.

23 November 1813. James Bodmer. No 2,755

Describes a breech-loading gun intended for all ordnance except mortars. The breech is closed by a conical plug which is held in place by a wedge. The touch hole is in the conical plug and has a valve which is opened by the priming tube but closed when the gun fires.

There were two other patents which dealt with the firing of guns. These were Forsyth's of 11 April 1807, No 3,022, which was to revolutionise firearms generally from that date on (cannons being affected only in the 1840s), and Fox's of 15 January 1820, No 4,427, which describes detonating powder in a priming tube fired by percussion.

APPENDIX 2
ORDNANCE OF THE ROYAL ARTILLERY, 1790–1820

Nature	Length ft in		Weight cwt	Remarks
GUNS				
Brass				
42 pdr	10	6	66	Obsolete by 1816 but little used after 1800
to	8	0	52	
32 pdr	10	0	55¼	Obsolete by 1816 but little used after 1800
to	9	6	42	
24 pdr	9	6	53	
	8	0	41¾	Out of service 1811 owing to failure in the Peninsula War
	6	3	24	
	5	0	16¾	
18 pdr	5	9	18	Out of service 1811 owing to failure in the Peninsula War
12 pdr	9	0	31½	
	7	6	22¼	These three lengths were out of use by 1800
	6	6	21¾	
	6	6	18	Medium 12 pdr. Considered excellent
	5	0	8¾	Obsolete by 1800
	5	0	12	Light 12 pdr used in Canada and countries of difficult terrain
9 pdr	6	0	13½	Although first cast in 1719, not used from 1750 to 1808. After 1808 a much used field gun
6 pdr	8	0	19¼	Obsolete by 1800
	7	0	12	Long 6 pdr. Good shooter but not a good travelling gun

Aug 1863. Volunteers Authorised to use
Martello 28 Enchantress Tower for a
Practice Range

Occupied by Regular R.A. Gunner
with instructions to call on Volunteers
to work gun in event of invasion poss.
up to 1877 32 Pdr s.m.L.

6.3.66.. 37. Martello Cliff End Dismantled
Request to assist
3 wagon loads to Folkstone Gun 56 cwt
Platform 33 cwt Carriage 140 cwt

Shot. 2 tons 3 cwt. & ammunition etc 14 cwt.
Z. 1. C Coast Brigade R. A. Folkstone.
letter from above.

REFRESHMENT
PERIODS

FORM G. 12/2

SUSSEX CONSTABULARY

Statement Form

NAME ————————————————————

AGE ———————— DATE OF BIRTH ———————— OCCUPATION ————————

ADDRESS ————————————————————

PLACE TAKEN ————————————————————

TIME & DATE
COMMENCED

TIME & DATE
COMPLETED

1873 Chemical Works ex Castle Estate
Date in year erected precluding use of target
area.

14.6.73. Final Match Rye Battery & 93 other detachments with 40 Pounder Armstrong.
All detachments missed Guns commanded
by Sgts & in one case a Corporal

Armstrong 40 Pdr 10'. 1" long 35 cwt
Service Charge 5 lbs. Projectile 41 lb.

Nature	Length		Weight	Remarks
	ft	in	cwt	
	6	0	8¾	Obsolete by 1800
	5	6	8	Obsolete by 1795
	5	0	5½	Light 6 pdr
	4	6	5	Obsolete by 1800
3 pdr	7	0	11¾	Obsolete by 1800
	6	0	6	Heavy or long 3 pdr
	3	6	2½	Light 3 pdr. Used later as the gun of the mountain service
	3	0	1¾	Light Infantry gun. Obsolete 1795
	3	0	1½	Obsolete by 1800
1 pdr	5	0	2½	Obsolete by 1815 but used for a time afterwards in the colonial service
	6	0	3	Obsolete before 1815
	7	0	3¼	Obsolete before 1815

Iron

Nature	Length		Weight	Remarks
42 pdr	10	0	67	Mainly used in garrisons and coastal artillery
	9	6	65	
32 pdr	10	0	58	Used on fortified and battering trains, siege train and coastal artillery
	9	6	55	
24 pdr	10	0	52	Used on siege trains
	9	6	49½	
	9	0	47½	
18 pdr	9	6	42	Garrisons, siege trains and heavy field batteries
	9	0	40	
12 pdr	9	6	34	Garrison and siege trains but after 1800 relegated to coastal artillery
	9	0	32	
	8	6	31½	
	7	6	29¼	
9 pdr	7	6	24½	Garrison service and after 1800 relegated to coastal artillery
	7	0	23	
6 pdr	8	0	22	Used in garrisons but relegated to coastal artillery after 1800
	6	0	16½	
4 pdr	6	0	22¼	Obsolete by 1795
	5	6	11¼	
3 pdr	4	6	7¼	Obsolete by 1800

In most of the above categories there were small variations in length and weight.

For Carronades see tables on p 87.

HOWITZERS
Brass

Nature	Length		Weight	Remarks
10 in	3	11½	25¾	Obsolete in 1811 after failure in the Peninsula War, but cast until 1816
8 in	3	1	12¾	Out of service in 1811 but cast until 1820
Heavy				
5½ in	2	9	10	Also used in garrisons
Light				
5½ in	2	2¾	4	Discarded after 1820 (last cast in 1819)
4⅖ in	1	10	2½	After 1820 used for colonial service and with mountain artillery; cast until 1859

Iron

Nature	Length		Weight	Remarks
10 in	5	0	39½	In service by 1820. Used in battering trains
8 in	4	0	20½	In service by 1820. Used in garrisons

Nature	Length ft in	Weight cwt	Remarks
MORTARS			
Brass			
13 in	3 7½	25	Obsolete about 1820
10 in	2 9	10¼	
8 in	2 1¼	4¼	
5½ in	1 4¼	1	'Royal' Mortar
4⅖ in	1 1½	¾	'Coehorn' Mortar
Iron			
13 in	3 7½	36	
10 in	2 9	16	
8 in	2 1	8	

Iron stone mortars (firing bags of stones) up to 13in calibre were used. Their length was 1ft 4in and their weight 10 cwt.

For rockets in use by the Royal Artillery, see table on p 36.

APPENDIX 3
ORDNANCE OF THE ROYAL NAVY, 1790 AND 1820

1790

Nature	Length ft in	Weight cwt	Remarks
GUNS			
42 pdr	9 6	65	Not a great number of these were in use at this time
32 pdr	10 0	50 ⎤	
	9 6	55 ⎦	This calibre of gun was greatly used in the Royal Navy
24 pdr	10 0	52	
	9 6	49	Used on main and middle decks
	9 0	47	Upper deck of 2nd rate and lower of some 4th rate ships
18 pdr	9 6	42 ⎤	Used as upper deck guns on 74 gun ships, main decks of
	9 0	47 ⎦	frigates, and on other ships
12 pdr	9 6	34 ⎤	
	9 0	32	
	8 6	31½	Greatly used in ships of the line and as chase guns
	8 0	28½ ⎦	
9 pdr	9 6	30¼ ⎤	
	9 0	29	
	8 6	27½	Occasionally used as broadside guns and also as chase guns
	8 0	26½	on frigates
	7 6	24½	
	7 0	23 ⎦	
6 pdr	9 0	24	
	8 6	23	
	8 0	22	
	7 6	20½	Chase guns of frigates
	7 0	19	
	6 6	18	
	6 0	16½	Chase guns of sloops
4 pdr	6 0	12¼	
	5 6	11¼	
3 pdr	4 6	7¼	
½ pdr swivel	3 0	1½	

Nature	Length		Weight	Remarks
	ft	in	cwt	
MORTARS				
Brass				
13 in	5	3	82	Obsolete in 1813
10 in	4	8	33	Obsolete in 1813
Iron				
13 in	5	3	82$\frac{1}{4}$	
10 in	4	8	41	
CARRONADES				
68 pdr	5	2	35	Before the introduction of 68 pdr guns after 1820, two carronades were placed on the lower deck
	4	0	29	
42 pdr	4	4	22$\frac{1}{4}$	Upper deck, fourth rate ships
32 pdr	4	0	17	General use for arming quarter decks and forecastle as well as poop decks
24 pdr	3	8	13	Not much used in the Royal Navy except for boats
	3	0	11$\frac{1}{2}$	
18 pdr	3	3	10	The main decks of small sloops and in boats
	2	4	8$\frac{1}{4}$	
12 pdr	2	2	5$\frac{3}{4}$	Some cutters and boats
6 pdr	2	8	4$\frac{3}{4}$	Obsolete by 1813 but used after this date on King's and Revenue cutters

1820

Nature	Length		Weight	Remarks
GUNS	ft	in	cwt	
42 pdr	9	0	67	A reduction in length since 1790
32 pdr	9	6	55$\frac{1}{2}$	Reduction in length since 1790
	8	0	49$\frac{3}{4}$	
24 pdr	9	6	50	An increase in the number of lengths since 1790
	9	0	47$\frac{1}{2}$	
	8	0	43$\frac{1}{2}$	
	7	6	40	
	6	6	33	
	6	0	31	
18 pdr	9	0	42	A reduction in length and an increase in number of types
	8	0	37$\frac{1}{2}$	
	6	0	27	
12 pdr	9	0	34	A reduction in length and number of types
	8	6	33	
	7	6	29$\frac{1}{2}$	
9 pdr	9	0	31$\frac{1}{2}$	A reduction in length and number of types
	8	6	28$\frac{1}{2}$	
	8	0	26$\frac{1}{2}$	
	7	6	26	
	7	0	25	
6 pdr	8	6	23	
	8	0	22	
	7	6	20$\frac{1}{2}$	
	7	0	20	
	6	6	18	
	6	0	16$\frac{1}{2}$	

4 pdr, 3 pdr and $\frac{1}{2}$ pdr swivel as in 1790.

Mortars and carronades in use in 1820 are the same as those listed in tables above.

APPENDIX 4
TABLES OF WEIGHTS AND DIMENSIONS OF CARRIAGES, BEDS, WHEELS AND TRUCKS, 1790–1820

Weight of carriages for iron guns and howitzers, and of beds for mortars

IRON GUNS

Pdr	Travelling		Garrison	
	gun carriage	limber	wood	iron
	cwt	cwt	cwt	cwt
42	—	—	16½	29
32	—	—	15	22½
24	24½	7½	13½	20½
18	17½	7½	12½	17½
12	16½	7½	11½	16
12	12	6	—	—
9	11½	6	10½	14½
6	—	—	9	15½
4	—	—	7	—
3	—	—	6½	—

HOWITZERS

10 in	29½	7½	19½	
8 in	21	7½	16½	

MORTARS

	Mortar bed		
	sea	land	iron
	cwt	cwt	cwt
13 in	39	21½	50
10 in	32	10	23
8 in	—	6	12
5½ in	—	1¼	—
4⅖ in	—	1	—

Iron garrison carriages were used in time of peace and in warm climates, also in well-covered positions.

Weight of carronade block trail garrison carriage

Nature of ordnance	Wood			Iron		
	cwt	qr	lb	cwt	qr	lb
68 pdr	17	1	25	—	—	—
42 pdr	17	1	25	—	—	—
32 pdr	7	1	4	11	3	9
24 pdr	7	3	21	10	3	—
18 pdr	6	3	20	10	3	—
12 pdr	6	1	—	8	—	—

Weight of field carriages

Nature of carriage		No of rounds	Weight			Total			Remarks
			Cwt	Qr	Lb	Cwt	Qr	Lb	
12 pdr	Gun	—	18	—	—				

Nature of carriage		No of rounds	Weight Cwt	Qr	Lb	Total Cwt	Qr	Lb	Remarks
	Carriage	—	12	3	9	42	I	14	
	Limber	—	8	3	6				
	Stores	18	2	2	27				
12 pdr ammunition waggon	Body	—	11	—	5				Weight of each
	Limber					30	I	2	each
	Ammunition	72	10	3	20				Charge 17 lb
12 pdr spare wheel carriage	Gun	—	13	I	21				
	Carriage	—	8	2	11	31	—	21	
	Stores	—	9	—	17				
9 pdr	Gun	—	13	2	—				
	Carriage	—	11	3	6	37	—	19	
	Limber	—	8	—	17				
	Stores	32	3	2	24				
9 pdr ammunition waggon	Body spare wheel	—	11	—	6				Weight of each
	Limber	—	8	—	24	29	—	2	each
	Stores	84	9	3	—				Charge 13 lb
9 pdr spare wheel carriage	Carriage	—	12	2	—				
	Limber	—	7	3	11	29	2	5	
	Spare stores	—	9	—	22				
Heavy 6 pdr	Gun	—	12	I	16				
	Carriage	—	11	—	24	35	I	7	
	Limber	—	8	—	—				
	Stores	50	3	3	5				
Heavy 6 pdr ammunition waggon	Body spare wheel	—	11	—	6				Weight of each
	Limber	—	8	—	24	29	3	16	each
	Stores	140	10	2	14				Charge 8½ lb
Heavy 6 pdr spare wheel carriage	Carriage	—	11	2	4				
	Limber	—	7	3	11	28	2	9	
	Spare stores	—	9	—	22				
Light 6 pdr	Gun	—	6	—	—				
	Carriage	—	8	3	6	25	2	23	
	Limber	—	8	—	5				
	Stores	40	2	3	12				
Light 6 pdr ammunition waggon	Body spare wheel	—	10	2	22				Weight of each
	Limber	—	8	—	5	28	—	3	each
	Ammunition	130	9	I	4				Charge 8 lb
Light 6 pdr spare wheel carriage	Carriage	—	9	2	11				
	Limber	—	7	3	11	36	2	16	
	Spare stores	—	9	—	22				
Heavy 3 pdr	Gun	—	6	—	—				
	Carriage	—	8	I	24	25	I	18	
	Limber	—	8	—	6				Weight of
	Ammunition	72	2	3	16				each
Heavy 6 pdr ammunition waggon	Body spare wheel	—	10	2	19				Charge 4½ lb
	Limber	—	8	—	6	21	2	13	
	Ammunition	170	2	3	16				

Nature of carriage		No of rounds	Weight			Total			Remarks
			Cwt	Qr	Lb	Cwt	Qr	Lb	
Heavy 3 pdr spare wheel carriage	Carriage	—	9	—	15				
	Limber	—	7	3	11	24	3	1	
	Spare stores	—	7	3	3				
Light 3 pdr	Gun	—	3	—	—				
	Carriage	—	4	—	3	11	1	20¼	Weight of each charge 3 lb 14 oz
	Limber	—	3	1	13				
	Ammunition	30	1	—	4¼				
Limber	Cart	—	3	6	14	5	3	22½	
	Ammunition	60	2	—	—				
3 pdr mountain service gun drawn by two mules	Gun	—	2	1	3				
	Carriage	—	2	1	4	6	3	13	
	4 small boxes	—	—	2	18				
	Ammunition	48	1	2	16				
When carried on the back of two mules	2 large boxes	—	—	2	1	2	1	1	
	Ammunition	45	1	3					
Heavy 5½ inch howitzer	Howitzer	—	10						
	Carriage	—	12	3	5	34	3	6	
	Limber	—	8	1	3				Weight of each charge 19 lb
	Ammunition	22	3	2	26				
Heavy 5½ inch howitzer ammunition waggon	Body spare wheel	—	11	—	7				
	Limber	—	8	1	3	29	—	20	
	Ammunition	58	9	3	10				
Light 5½ inch howitzer	Howitzer	—	4	3	—				
	Carriage	—	10	1	23	26	3	21	
	Limber	—	8	1	16				Weight of each charge 17 lb
	Ammunition	22	3	1	10				
Light 5½ inch howitzer ammunition waggon	Body spare wheel	—	10	3	3				
	Limber	—	8	—	18	28	2	15	
	Ammunition	64	9	2	24				
4⅖ inch howitzer	Howitzer	—	2	2	—				
	Carriage	—	5	1	2	12	1	14	
	Limber	—	3	1	18				Weight of each Charge 8½ lb
	Ammunition	16	1	—	24				
Limber 4⅖ inch	Cart	—	3	2	20	6	—	12	
	Ammunition	32	1	—	24				
Forge waggon with lockers etc	Body	—	9	—	8				The new bellows proposed will be 1 qr 26 lb
	Limber	—	7	3	11				
	Bellows	—	—	2	15	19	—	24	
	Anvil		1	2	8				
	Smith's		2	—	5	4	—	5	
	iron tools		2	—	—				
Flander's waggon Store limber carriage	Ammunition		15	—	2	15	—	2	
	Body and spare wheel		10	—	11				
	Limber		7	3	11	17	3	22	

Nature of carriage		No of rounds	Weight			Total			Remarks
			Cwt	Qr	Lb	Cwt	Qr	Lb	
Ball cartridge waggon new pattern Gen. Millar	Body		7	—	26				Fall for sling cart 3 inch rope, 12 fathom sling, 4 fathom length of lever for sling cart 8 ft
	Limber		7	2	4				
	20 boxes								
	79½ lb		1	2	22	34	—	18	
	20,000 rounds		17	1	18				
	1,000 flints		—	1	4				
Sling cart			16	2	2	16	2	2	Length of common handspike 6 ft
Large gyn	Gyn		8	2	16				
Common gyn			2	3	25	11	2	13	Length of cheek for a large gyn for traversing platform 18 ft 6 in
	Platform		6	—	20				
Traversing			2	2	19	8	3	11	
	Carriage								
Platform carriage	Complete		50	—	—	50	—	—	Length of cheek for a common gyn 16 ft
	Small do.		21	—	23	21	—	23	
Large devil carriage	Large		27	2	23	27	2	23	
	Small		7	2	8	7	2	8	
French carriage	Carriage		13	2	18				Windlass 6 ft
	Boat		5	1	10	13	2	18	
Large pontoon	Gyn,		15	3	21				
	blocks, etc		10	—	—	5	1	10	
Carriage	Appurtenances		17	—	11	43	—	4	
Small do.	Carriage		12	—	6				
	Boat		7	—	—	28	3	18	
	Appurtenances		9	3	12				
Small sea boat taken	Carriage		7	—	5				
with pontoon bridge	Wood boats		3	2	—	10	2	5	
Ball cartridge cart			7	3	4				
Baggage cart			8	3	22				
Hand cart			4	1	25				
Trench cart			2	2	13				
Tumbril cart			7	—	22				

Dimensions and weights of wheels

Class	ft	in	cwt	qr	lb
1st	5	—	4	—	26
2nd	5	—	3	2	1
3rd	4	4	2	—	15
4th	3	—	—	3	19
Platform carriage					
Hind	5	—	5	2	7
Fore	4	—	4	2	2
Devil's carriage	7	—	11	—	24
Sling cart	5	6	—	1	3
Flanders waggon					
Hind	5	—	3	2	1
Fore	4	3	2	3	20
Hand cart	4	4	2	—	15

Dimension of trucks for sea and garrison carriages
(Sea trucks in wood, land trucks in iron)

Nature of ordnance	diameter of truck (inches)		thickness of truck (inches)
	fore	hind	fore and hind
42 pdr	19	16	6·5
32 pdr	19	16	6·0
24 pdr	18	15	5·5
18 pdr	18	14	5·0
12 pdr	16	14	4·5
9 pdr	16	14	4·0
6 pdr	14	12	3·5
3 pdr	14	10	3·0

APPENDIX 5
UNIFORMS OF THE ROYAL ARTILLERY
1790–1820

Royal Artillery

The first pattern of head-dress worn by privates in this period was introduced by General Order of 29 June 1785. The hat, which was worn only by privates and non-commissioned officers, was conical in shape with a narrow brim and a broad band of yellow lace around the base. On the top front of the hat was a leather cockade, behind which was fitted a red plume. The civilian drivers appear to have worn a similar-style hat. The officers wore a fur-crested helmet with a cockade and plume on the left side. These styles were discontinued (in about 1797) in favour of the cocked hat for both officers and men.

The uniform consisted of a blue coat with red cuffs, collar and turned-back lapels with ten buttons each side which fastened at the neck and sloped away showing the white waistcoat. Around the waist was tied a crimson sash with the tassels hanging down on the left side. The uniform for the privates was of the same design but of poorer quality material. White breeches were worn with black gaiters reaching to the knee, buttoned on the outside.

Two white buff leather cross-belts were worn by privates, one over the left shoulder suspending the ammunition pouch on the right hip, the other suspending the bayonet in a frog on the left hip. The pouch was in white buff leather in accordance with King's Orders of 1771. The pouch had a large flap on which was placed a brass badge which was the crown with a semi-circular scroll beneath with the inscription 'R.Artil.' followed by the battalion number. In the centre of the cross-belt was a red flask cord suspending a small priming horn above the pouch. This is clearly shown in the water-colour by Edward Dayes. On the cross-belt were two pockets which held a brass-headed hammer and a pricker which was used for cleaning the vent and

piercing the flannel bags of the cartridge to allow the flame from the lit priming powder to ignite the main charge.

Haversacks were also carried by the marching battalions. These are also shown in the Dayes water-colour and were made of grey canvas with a painted circle in the centre featuring the crown in heraldic colours above the letters 'R.A.'; beneath this was the battalion number followed by the inscription 'Batt'.

Lieutenants and captains wore one gold epaulette on the right shoulder, and field officers wore a pair. Staff sergeants wore two bullion epaulettes, sergeants wore two gold lace straps, corporals wore red worsted fringe epaulettes, bombardiers wore one fringe epaulette, and privates wore red worsted straps.

Men of the invalid battalions (composed of pensioned Artillerymen) wore the same uniform as the marching battalions, except that the coats were lined with red, not white, and their breeches and waistcoats were blue.

An order of 12 October 1796 stated that the Artillery should conform to the regulation hat and sword of the Infantry, which involved an issue of a cocked hat.

In about 1799 a new uniform was adopted by officers and men. The officers now wore a blue double-breasted coat with the leading edge piped in scarlet. The collar and cuffs were scarlet but with no adornment at all except for buttons. Occasionally the top of the coat was left open and buttoned back so that it showed the scarlet cloth lining. The crimson waist sash was knotted on the left side with the tassels hanging down. White breeches with black leather boots continued to be worn.

The other ranks wore a short-tailed blue coat with red collar and cuffs edged in yellow worsted

58 Royal Artillery uniforms. Left, *officer, 1797;* right, *private, 1797. Watercolours by Edward Dayes, drawn for* HRH *The Duke of York*

tape. On the cuffs were woven button loops in red cord. The shoulder straps were red with yellow tape edging and worsted tufts. The coat was single-breasted and the front was ornamented with rows of bastion-ended yellow tape. The rest of the uniform and equipment remained the same.

In 1806, the rank and file were ordered to adopt the shako worn by the Infantry of the line, and in April 1808 General Orders laid down that the plume worn by officers and all ranks should be the same length in the marching battalions.

General Orders of 14 January 1812 changed the uniform for officers, bringing it more into line with that worn by Infantry Officers. The coat was blue with scarlet collar and cuffs, with gold lace gorget tabs on the collar and four buttons with gold lace button loops on the cuffs. The lapels were turned back forming a plastron front, ornamented with rows of lace each side and fastened by hooks and eyes. The epaulette(s) worn conformed to the order of 2 March which laid down that field officers and captains should have bullion and lieutenants, fringe. (This referred to the quality of the tassels that hung from the point of the shoulder.) The shako, known as the Wellington or Belgic shako, was introduced on 24 December 1811. The body and false front were made in black beaver for

officers and felt for rank and file. The lines were white cord and were fitted to the shako at each side. The plate was a crowned oval with baroque edges with the garter in the centre with the G.R. cypher within it. Below this was a mortar with two flaming grenades.

The uniform of the other ranks remained the same except that the button loops were now of yellow worsted tape and not in crimson cord. This uniform is well shown in the print reproduced on p. 104.

After Waterloo, the Belgic shako was replaced by the bell-shaped 'Regency' shako. At this time white breeches with knee gaiters were abolished and grey trousers substituted. The officer's coat lost all its gold lace, except for the button loops on the cuffs.

Royal Horse Artillery
On the introduction of the Horse Artillery into the Army in 1793, there were two troops, A and B.

The helmet worn by the officers of both troops was the Light Dragoon pattern with a crimson turban tied in a rosette at the back. The helmet had a bearskin comb on the top, and across the front above the peak was a band in gilt metal on which was the title 'Royal Horse Artillery'. On the right side of the helmet was a gilt badge and on the left, fitted into a socket behind the turban, was a white feather plume.

The coat was blue with scarlet collar, cuffs, turnbacks and lapels. It fastened at the neck and sloped away showing the white waistcoat worn beneath. The shoulders were adorned with epaulettes of interwoven gilt rings on a backing of scarlet cloth. A crimson sash was tied round the waist and knotted on the left side. A buff leather cross-belt was worn over the left shoulder attached to a pouch worn in the middle of the back. White breeches and black boots completed the uniform.

The uniform worn by the other ranks was similar but without the crimson waist sash.

In 1799 the uniform was completely changed to conform more with that of the Light Dragoons. The same helmet continued in use but the coat gave way to a shell jacket with a fully frogged front in gold for the officers and in yellow for the rank and file. A crimson sash was worn round the bottom of the shell jacket but about three inches above the natural waist, as was the fashion. Amongst the officers, there appears to have been a certain latitude in the number of gold bars on the front of the jacket, according to General Mercer's *Military Reminiscences of the late 18th and early 19th centuries*.

A jacket in the Royal Artillery Institution has twenty-four bars of lace and seventy-two gilt ball buttons. A new pattern of helmet was introduced after the turn of the century which had a more pleasing shape, dipping down at the back, and shin scales were fitted each side by rosettes. The comb was enlarged and General Mercer states that the battery collar-maker improved the combs by combining the new combs with the combs of the old helmets. This pattern of helmet was worn by the Royal Horse Artillery until 1827 and is well shown in the Hamilton-Smith print of Rocket Troop.

The other ranks' uniform was the same except that a white buff cross-belt with two sword slings was worn and a holster in brown leather was attached to the waistbelt on the left side. The overalls were grey with a broad red stripe down the outside seam of each leg. Officers wore these overalls on active service.

The uniform of the Royal Horse Artillery remained virtually the same until 1828. In 1804 the length of the jacket was slightly altered, but a return was made to the original pattern in 1810. Mention must be made here of the newly formed Rocket Troop, who appear to have worn the same uniform as the privates of the other troops but without the cross-belt suspending the sword. In its place, but over the other shoulder, they wore a white buff cross-belt and pouch, the sword being carried on the saddle.

The black leather linings of the overalls gave way to brown leather in about 1811 when the height of the leather round the bottom of the legs was reduced to three inches to protect the overalls from the spurs.

Civilian drivers; Corps of Captains, Commissaires and Drivers; Corps of Drivers; 1790–1820

Until 1794, drivers were supplied by the civilian contractors who supplied the horses for the Artillery. The drivers were supplied with a uniform of a coat, white breeches and black boots. On the left arm they wore a blue band with the initials G.R. in yellow. The waggon-master was always a Royal Artillery officer.

In 1796 this system was abolished and a permanent corps of Captains, Commissaires and Drivers was established to take over the work carried out by civilian drivers. They were supplied with the tall tapering hat worn by the marching battalions and a white smock, breeches and knee-boots. Foul-weather coats made of leather were issued in inclement weather. The officer's uniform was that of the Royal Artillery.

In 1796 a new uniform was adopted which was very similar to that worn by the troops of Horse Artillery. The Light Dragoon helmet had a dark blue turban and the shell jacket was blue with red collar and cuffs with yellow cord frogging adorning the front. In 1806, this uniform was superseded by a new pattern, possibly under pressure from the Royal Horse Artillery who considered the drivers' uniform too similar to theirs. The shell jacket lost all the yellow frogging on the front and instead had the addition of a row of fifteen buttons each side of the central row, giving the appearance of a plastron shape. White breeches and black riding boots completed the uniform.

In 1810 yet another change was made in the uniform. The helmet was retained but the blue coat with its red collar and cuffs lost the two extra rows of buttons on the front and was ornamented instead with rows of yellow bastion-ended tape, similar to that on the coats of the marching battalions. A cross-belt was worn over the right shoulder as well as a white strap suspending a canvas haversack

59 *Rocket troop, Royal Horse Artillery. From an aquatint by Stadler drawn by Hamilton Smith, 1815*

on the left hip. Blue grey overalls with a broad red stripe on the outside of each leg were issued. These had buttons on the red stripes and leather linings to the legs ending in a false boot at the bottom of each leg. These were termed booted overalls.

Before the Waterloo campaign, the uniform had been changed again. The coat reverted to the style worn before 1810, except that the three rows of buttons were lined each side in red piping which joined across the top of the chest forming a plastron shape. The cross-belt was done away with, and a white buff waist-belt with snake fastening was substituted. This was the uniform worn until 1822, when the Driver Corps was merged with the Royal Artillery, all engagements then being as gunner and driver.

FORAGE AND UNDRESS CAPS

1790, Foot Artillery.   Black leather with G.R. badge in brass.

1793, Horse Artillery.   The same as for the foot artillery, but always kept well polished and clean, whereas those of the foot artillery were never cleaned, nor meant to be. Used for undress parades.

For stables etc, the Royal Horse Artillery, according to Mercer, wore a blue cloth cap edged in red and tied by laces at the back. These undress and forage caps were worn throughout the period 1790–1820.

BUTTONS

1785–1802   Three cannon balls in line at the top and three cannons below in line, raised in a Norman shield. The buttons were convex. Other ranks' buttons were almost flat and of bronze but later were made hollow-backed. The Royal Horse Artillery wore a ball button with this device.

1802–20   On the button was a garter surmounted by a crown with the G.R. cypher inside. In the garter belt were the words 'Royal Regiment of Artillery'. Officers' buttons were gilt and convex while the buttons for the other ranks were in bronze. The Royal Horse Artillery changed from this pattern in 1808.

1808–55   Royal Horse Artillery had engraved ball buttons with the garter and cypher surmounted by a crown, as in the 1802 button, but with different wording in the garter belt. The following variations have been noted:

Royal Horse Artillery
Royal Horse Artil.
Royal Horse Arty.

Different patterns of buttons were worn by the Irish Artillery, who did not become part of the Royal Artillery until 1801. One design was a shield with a crowned harp surmounting a gun with a cannonball each side of the crown. Another design known is that of a shield bearing the Ordnance arms surmounted by the crowned harp and with the word 'Artillery' on the left of the shield and 'Royal Irish' on the right.

SHOULDER-BELT PLATES

Permission to wear shoulder-belt plates was granted in 1779 to certain of the battalions. General Order of 30 October 1796 stated that the Master-General was extremely willing to comply with the wishes of the Colonels Commandant and desired that they fix a cross-belt plate for the sword. There are a great variety of plates attributed to the Royal Artillery for this period. Described below are some worn at this time.

1790   Oval plate in brass with the shield bearing the Ordnance arms as on the buttons. The word 'Royal' was engraved above the shield and the word 'Artillery' below.

1790–1815   Oval bronze plate for other ranks engraved with the crown above three cannon balls in a line, below which were three guns one above the other. There was a scroll above the crown engraved 'Royal British' and another below engraved 'Artillery'. This plate is thought to have been so engraved from June 1812 to distinguish the British Artillery from the Royal German Artillery. On the other hand it could be engraved with the word British to distinguish it from the Irish Artillery prior to 1801.

1790–1810   The 3rd battalion wore a rectangular belt-plate with a roped edge. In the centre was an engraved gun surmounted by the crown with GIII on one side and R on the other. On each side of the gun was a rammer and portfire stick. Beneath the gun was the figure '3' flanked by two piles of shot. Engraved at the bottom of the plate were the words 'Royal Artillery'.

SHAKO PLATES

1804–12   The plate worn on the shako was die-struck, with, in the centre, a crowned garter with a G.R. cypher within it, the garter inscribed with the words 'Royal Regiment of Artillery'. The garter was superimposed on a trophy of flags, arms, etc. Below the garter was a mortar on its bed flanked by piles of shot.

1812–16   The plate was a die-struck crowned oval with baroque edges. In the centre was a garter belt inscribed 'Royal Regiment of Artillery' with the G.R. cypher inside. Beneath this was a mortar on its bed with two flaming grenades.

1816–20   The other ranks' plate for the Regency shako is thought to have been a crowned circle with a roped edge bearing the badge of three guns one above the other in the centre.

APPENDIX 6
UNIFORMS OF THE ROYAL NAVY,
1790–1820
(THE OFFICERS AND MEN CONCERNED
WITH THE GUNNERY)

The full dress uniform of a lieutenant, introduced in 1787, was a blue tailed coat which was double breasted but usually worn open. Worn in this fashion, the white lining was shown and formed the lapels which were buttoned down with nine gilt buttons on each side. The high collar, also worn open to reveal the stock, was blue with a gilt button and woven buttonhole on each side. Stocks tended to be white rather than black about 1800, but this was a matter of personal preference. The round cuffs were white with three gilt buttons in a line on each. The tails of the coat were plain, with two pockets and pocket flaps decorated with gilt buttons. A white waistcoat was worn under the coat and left open at the top to show the shirt ruffle, although the shirt styles were left to the individual. There were fifteen buttons on the waistcoat. (The numbers of buttons on coats and waistcoats during this period often varied and the figures quoted are an average.) White pantaloons with white stockings and black buckled shoes completed the uniform.

The hat was worn across the head, and by 1787 the cock had become fairly small and the hat close. It was edged in navy pattern lace and decorated with a black silk rosette and button on the left side.

A black leather shoulder-belt, fastening on the chest with an oval gilt plate, suspended the sword on the left side.

This pattern of uniform was in use until 23 March 1812, when a modified pattern was introduced. The undress coat was blue with white piping.

In March 1812, lieutenants were authorised to wear a single plain gold-fringed epaulette on the right shoulder. In a contemporary portrait of an unidentified officer (in the National Maritime Museum, Greenwich), the coat is shown hooked up, giving the impression of a white plastron shape. The collar has a small gilt button and a woven buttonhole. Otherwise the uniform remained the same.

During the Napoleonic Wars, conditions were such that lieutenants found themselves in command of small vessels and needed a second-in-command. To fill this need, the rank of sub-lieutenant was created in 1804. Sub-lieutenants had no full dress uniform and wore only the undress uniform of the rank of lieutenant.

This uniform was similar in cut to the full dress with the collar, cuffs, lapels, pocket flaps and edge of the coat piped in white. After 1800 the hat was worn fore and aft and had the additional decoration of a loop of lace across the cockade.

The uniform worn by midshipmen in the corresponding period remained virtually unchanged. The coat was blue and single-breasted with nine gilt buttons down the front. It was sometimes worn open, in which case the white waistcoat could be seen. The coat tails were decorated with buttons in the pleats and on the pocket flaps. The cuffs had three large buttons on them with two smaller buttons at the side, one above the other. The use of the two small buttons was a new style copied from civilian dress of the period. The high blue collar had two white cloth patches each side decorated with a gilt button and woven buttonhole. White pantaloons with white stockings and buckled shoes completed the dress. The hat was bicorn, worn across the head, and decorated with a black silk rosette, a lace loop and a button. There were no regulations for hats at this period, but a uniform style with minor variations appears to have been in wear.

The undress uniform of a midshipman of the period is described in the pattern book of the military and naval tailors Welch and Stalker, and is as follows:

60   *Lieutenant, Royal Navy, 1787–1812*

Blue coat, single breasted. X [cross] flap with 3 holes . . . plain Blue round cuff with 3 buttons. Upright collar turn'd back with White and a Breast Button at the end. Anchor Buttons. No White Edges but lined with White Rattinet or Shalloon. White Cloth or Cassimere Waistcoat and Breeches, Single or Double breasted no skirt.

Towards the end of the 1780s, 'jacket suits' were mentioned as part of a midshipman's kit. These short-tailed coats were being worn throughout the Navy as a working dress after 1800.

At this period, officers' and midshipmen's dress was dictated not only by regulations and fashion but by necessity. Officers sometimes cut the tails off their uniform coats for easier movement on board ship and during active service. This was the ancestor of the later round jacket.

The dress worn by the Gunner was that prescribed for other warrant officers of the period. Warrant officers were given a uniform for the first time by an order dated 17 November 1787. The order stated that warrant officers were to wear:

Blue cloth coat, with blue lapels and round cuffs; fall down Collar; Three Buttons to the Pocket and Cuff; White Lining but not edged with white; Button with an Anchor, same as Captains former one:—White Cloth Waistcoat and Breeches.

In 1807 it was changed for masters and pursers but the Gunners wore the same uniform and did not adopt the crowned button of 1812 with which officers were to be distinguished.

The dress of the ordinary seamen was dictated by the bulk buying of materials by the purchasing authorities. The victualling yards supplied these 'slops' to the purser who, in turn, issued the material to the men, the cost of 'slop' clothing being deducted from the seamen's pay. The men were issued with twelve yards of duck and thread and needles. From contemporary accounts, it appears that, at the end of a week, all men were expected to be attired with some degree of uniformity.

The dress usually consisted of undyed off-white canvas pantaloons, flared at the bottom, and a blouse of the same material. The hair was tarred, although this practice was less common after 1800, and a scarf was worn to prevent the hair queue from rubbing the tar on the white blouse. Tarred straw hats, again made by the men themselves, were sometimes worn, although this was far from general. For shore or walking out, when allowed for those fortunate enough, a short blue jacket was worn, although this again differed in style as there

was no regulation to govern the design.

The dress of the seamen depended entirely on the cloth provided, and sometimes the striped trousers, in which men of the period have been popularly portrayed in films, were possible and were worn, but the normal wear appears to have been white. An order, No 38, of the ship *Pylades*, under the command of Captain Roberts, stated:

As the ship's company will have a better appearance by preserving a uniformity of dress, they are to be discouraged from purchasing any other clothing than blue, white or red waistcoats, blue or white trousers, black handkerchiefs and hats.

The seaman had no weapons that he carried on his person at all times, pikes, cutlasses and pistols being issued in action or for landing parties. These weapons were kept in ready-use racks on the gun-decks.

There were no regulations governing the shoulder-belt or cross-belt plates worn in the Royal Navy, but these were normally oval and made of gilt brass. They bore either the design of a foul anchor or the ship's name or a device of some kind. At times a waist-belt was worn in place of the shoulder-belt with a buckle in the shape of a serpent with a lion's head at each side. The cross-belt, however, seems to have been discontinued around 1812 in favour of a waist-belt of white buff leather with an oval plate bearing the design of a crown above a foul anchor. No belt regulations were issued, and belts were often of black leather instead of the white buff, and could be worn under or over the waistcoat. In some cases the belt was even worn over the coat, but this was very unusual.

Buttons worn by lieutenants were the same as those worn by captains, these being round and made of gilt brass with a roped edge. On the button was an inner oval of rope containing a foul anchor on a matted background. Buttons worn by midshipmen were those that were adopted by captains in 1774 and discarded in 1787 being plain and flat with a foul anchor on them. Midshipmen wore this design until 1812 when they adopted a button with a crown over a foul anchor, which remains, except for minor changes, the button of the Royal Navy today.

The button used by warrant officers was the same as that worn by captains between 1774 and 1787, being identical to that worn by midshipmen. Warrant officers were ordered to wear the captain's 1774 pattern button until as late as 1860, but many wore raised buttons with crowned foul anchor of the 1827 pattern from that date.

61   *Master, Royal Navy, 1787–1807*

APPENDIX 7
UNIFORMS OF THE ROYAL MARINE ARTILLERY, 1804–1820

On the formation of the Royal Marine Artillery in 1804 no specific uniform was ordered other than that prescribed for Royal Marines. A blue cloth undress fatigue jacket, however, seems to have unofficially crept into use. This jacket had red collar and cuffs after the style worn by the Royal Artillery. The use of this jacket prompted the captain of the bomb vessel *Hound* to complain to the Admiralty about this non-regulation jacket.

The Admiralty immediately wrote to Colonel Anderson of the Chatham Division and requested him to report 'upon what authority the dress of this party differs from the established Clothing of the Corps'. From Colonel Anderson's reply it appeared that the blue undress jacket with red collar and cuffs had been sanctioned by Lieutenant-Generals Campbell and Barclay. Captain Williams reported that in June 1805 he was

Induced to propose to Lieutenant-General Barclay . . . that on account of great injury which the uniform clothing of the Companies sustained from Powder, when at Gun and Mortar practice, but more especially the latter; they should be permitted to wear a blue jacket and a pair of overalls which they were to wear on all duties of fatigues and upon such occasions as might be necessary in order to preserve their uniform clothing. That in consequence of this proposal the Lieutenant-General in the orders for June 14th 1805 signified his approbation of this undress which was to be purchased at the expense of the men in lieu of the undress red jacket usually furnished to the Royal Marines.

Captain Williams goes on to say that the uniform jacket was examined at the annual inspection on 16 December 1805 by Lieutenant-General Campbell, who considered it an appropriate dress. The Secretary of the Admiralty replied that their Lordships had no objection to the fatigue jacket, but could not allow it to be worn when embarked.

In 1814, the Royal Marine Artillery other ranks were issued with grey trousers of the type issued to the Royal Artillery. In 1816, officers were at last permitted to wear the blue undress jacket and cap when employed on Artillery service. Prior to this date, officers had worn a red fatigue jacket, although the men, as has been explained, wore the blue.

At last, on 26 October 1816, Admiralty Instructions stated that the Royal Marine Artillery were to be dressed in the same way as the Royal Artillery, except that their buttons and hats were to be the pattern worn by the Royal Marines. On 19 December, a further order was issued stating that the head-dress was to differ from the Infantry Companies' by the addition of plate, cockade, tuft and band. The hat worn was similar in shape to that worn by the Royal Marines. It was a black, tarred-leather, tapering hat with a turned-up brim edged in white tape. The hat band was also in white tape. In the Royal Marines a cockade and plume were fitted on the top left side, with two black connecting cords to the brim on both sides. The Royal Marine Artillery wore their plumes in the top front of the hat behind a black rosette and with a brass plate beneath.

The hat plate for the other ranks was in stamped brass and of the shape worn on the stovepipe shako by the Royal Artillery between 1806 and 1812. On the plate there was the design of the crown surmounting a roped-edged garter belt with the words ROYAL MARINE ARTILLERY within the garter. The garter was superimposed on a trophy of flags, swords and muskets, which showed at each side. In the centre of the garter, on a horizontally lined ground, was a foul anchor. Beneath the garter was a mortar with a pile of cannonballs on each side.

The uniform accoutrements were virtually the same as those of the Royal Artillery except that Royal Marine Artillery officers wore the cross-belt plate and gorget of Royal Marine officers. Officers and other ranks of the Royal Marine Artillery wore the same buttons on their uniforms as their counterparts in the Royal Marines. The officers' buttons were gilt about 1800 with, in relief, the design of a foul anchor flanked by a wreath which joined beneath the anchor and the words ROYAL MARINES above the anchor. The other ranks' buttons had the same design, which was incised into the pewter button.

The belt-plate was oblong with the design of a lion on the crown. The gorget was in gilt and bore the design of the royal arms above a shield with the foul anchor in it. On each side of the shield was a spray of laurel.

62 *Officer and private, Royal Marine Artillery,*
1816. From a drawing by Captain J. S. Hicks,
RM

APPENDIX 8
SIDE ARMS OF THE ROYAL ARTILLERY,
1790–1820

Swords

Although there do not appear to be any specifications extant concerning the swords carried by the Royal Artillery in this period, there are numerous references and a number of portraits and drawings which give a good idea of the types of sword carried in the regiment.

Duncan's *History of the Royal Artillery* says that in 1782, amongst the equipment of a subaltern, there is one regimental sword, and Lieutenant-Colonel Anthony Farrington (who later became Lieutenant-General Anthony Farrington, a Colonel Commandant of the Regiment) in his Garrison Orders for 8 August 1794 stated that:

> The Regulation Sword has a straight blade and the length of it is as established by His Majesty's Regulations. It is to be worn with a Crimson and Gold Swordknot.

This would seem to refer to the sword which was carried by the majority of officers in the army and which eventually became the regulation pattern of 1796. It had a single-edged cut-and-thrust blade with a strong flat back some thirty-two inches from shoulder to point and one-and-one-eighth inches wide at the shoulder, and the hilt consisted of an urn-shaped pommel with a simple knucklebow and two shells, the inner of which was occasionally hinged to fold flat to the body, and the finial of the quillon was a large acorn. The grip was of wood, closely bound with twisted silver wire.

The Royal Horse Artillery of this period appear to have been equipped with two swords, as General Mercer in his *Military Reminiscences of the late 18th and early 19th centuries* states that:

> In the Horse Artillery besides the large regulation sabre, we had a small undress one so crooked as to be useless for anything else but a reap hook.

However, it would appear that the regulation sword for the Royal Horse Artillery was, in fact, the stirrup-hilted light cavalry sabre of the period for both officers and men, and this is borne out by many contemporary prints and pictures which show both officers and other ranks equipped with a steel-hilted sabre in a steel scabbard. Most of the blades of the officers' swords were decorated by being blued and gilt for half their length and engraved with the royal cypher, the royal coat of arms and other decoration. The engraving was filled with mercurial gold.

Other ranks of the Royal Artillery appear to have carried a brass-hilted hanger of a style which was common to a number of European armies. The solid cast grip and knucklebow was driven onto the tang of the slightly curved short blade and rivetted over. The scabbard was in black leather with brass mounts. This sword is shown in a print by Hamilton Smith entitled *Royal Artillery Privates 1815*. The privates were also armed with a carbine and bayonet.

Prior to 1750 Artillery sergeants had carried a linstock, but in that year they were ordered to carry instead a crossbar pike or spontoon. (Although this weapon was abolished for Infantry sergeants by General Order of 31 July 1830, it continued to be carried by Artillery sergeants until 1845 when it was replaced by a sword.)

The spontoon had a wooden shaft and a three-piece head. The double-edged spear point screwed into a socket which was attached to the wooden shaft with long langets and screws. The crossbar went between the blade and the top of the langets.

Muskets and pistols

The Royal Artillery privates carried a 37-inch-barrelled flintlock carbine which was 0·65 bore. This weapon weighed 7 lb 12 oz and was fitted with a triangular sectioned blade socket bayonet of 13 inches.

In September 1757 the Board of Ordnance had decided to keep a stock of 50,000 carbines with bayonets for Artillery and Highlanders at the Tower of London. The first pattern had a wooden ramrod, but this was changed to steel in 1772 at a cost of 2s 3d per carbine. The weapons were of a typical military quality with a steel barrel, brass furniture and a Tower-engraved lock. The woodwork was fitted with two swivels to take the white buff leather sling. The polished steel bayonet was carried in a brass-mounted leather scabbard, the top mount having a stud for fixing into the frog on the cross-belt. The cost of such a carbine in the late eighteenth century was about 24s. This continued as the standard weapon for Royal Artillery privates, except of course Royal Horse Artillery, right through the period, although a number of India pattern carbines were issued in 1813.

On their formation in 1793, the Royal Horse Artillery had a pistol designed and made specially for them. In January 1793 the Board of Ordnance asked Henry Nock, the leading government

contractor and one of the most famous English gunsmiths of his day, to design and supply eighty double-barrelled pistols. The Bill Books describe the pistol as:

> Pistol for Horse Artillery double barrelled, one of the barrels rifled 18 inches long, 13 balls to the pound . . . shifting butt [removable butt] and steel rammer.

The cost of each pistol was £8 and the weight was 7 lb 12 oz. Because of the size of this pistol, it is safe to say that it was used mostly with the butt on. The furniture of these pistols were brass.

These pistols seem to have been used by the Royal Horse Artillery until at least 1818, when

63 Royal Artillery privates standing by a gun mounted on a wooden garrison carriage. From an aquatint by Stadler drawn by Hamilton Smith, 1815

probably this and the normal cavalry pistol were both in use. Henry Nock died in 1805 and although it is on record that he made eighty pistols, presumably after his death others were made. His pistols may also have been replaced, when unserviceable, with the standard cavalry pistol, as prints of the period 1810–15 show a pistol carried in a short brown leather holster not long enough to take the 18 inch barrels of the double pistol.

APPENDIX 9
SIDE ARMS OF THE ROYAL NAVY,
1790–1820

Swords and dirks

Prior to 1805 there was no regulation pattern for naval officers' swords, but from about 1780 commissioned officers, and more particularly the senior officers, carried a sword of the following description.

The hilt had a knucklebow which had five small balls in the centre of the knuckleguard and on the obverse side of the shell, and a small openwork anchor in the centre of the shell. It had a fluted white ivory grip bisected by a gilt band with an oval shield on the outside, engraved with crown and foul anchor. The grip was surmounted by an octagonal pommel. The straight blade had a flat back and was fullered on both sides, one inch from the shoulder to the double spear point. The normal length was 32 inches, and the blade was an inch wide at the shoulder.

Some of these swords had blued and gilt decoration but the majority were plain, engraved with the royal coat of arms and royal cypher with patterns of foliage, etc.

The scabbard was of black leather with three gilt mounts, the top two bearing rings for suspension and the top mount also bearing a stud for wear in a frog.

The fighting sword of the period was normally a curved weapon with a flat back to the blade which terminated in a double spear point. The blade was about 28 inches long and $1\frac{1}{4}$–$1\frac{1}{2}$ inches wide at the shoulder. It was surmounted by a gilt stirrup hilt with two langets, which normally bore an engraved anchor. The pommel and backpiece were plain and the grip was usually knurled white ivory. The sword was carried in a black leather scabbard with three gilt mounts.

The first 'regulation' naval officer's sword came in 1805, but there is doubt as to the actual date of introduction, and the order of 4 August of that year which referred to the despatch of pattern swords to the port admirals can only be assumed to be the executive command for the adoption of the new pattern.

There were three types of weapon deposited with the port admirals. The first, for senior officers, had a lion mask and mane for the pommel and backpiece and a grip of carved ivory. The knuckleguard was a stirrup and had langets engraved with a foul anchor. The second, for lieutenants, was the same but with a grip of wood covered with black fishskin and bound with three gilt wires. Both patterns had straight blades 32 inches by one inch, engraved with the royal cypher and coat of arms. The senior officers' sword blade was invariably blued and gilt.

The third pattern was for masters, mates, midshipmen and warrant officers and had a plain backpiece and pommel with a black fishskin grip bound with gilt wires and a plain blade of the same dimensions as the other two patterns. All three swords were carried in black leather scabbards with three gilt mounts, the top one having a stud and a ring, the middle one solely a ring.

Flag officers carried the senior officers' weapon with the scabbard mounts engraved with a motive of oak leaves and acorns.

Although the dirk did not become official for midshipmen until 1856, several patterns of unofficial dirk were carried. One of these was of the same pattern as the officers' dress sword of the 1780s, having a gilt 4-sided pommel and a fluted ivory grip bisected by a gilt band bearing an oval shield engraved with a crown and foul anchor. There was no knuckleguard, but the cross-guard had a small open shell rimmed with five balls and containing an openwork foul anchor. The blade was about sixteen inches long, and the dirk was carried in a gilt-mounted black leather scabbard. A second type of dirk had a straight diamond-section blade approximately sixteen inches long with a straight crosspiece and a fluted ivory grip surmounted by a gilt cap. A third type had a curved blade and a lion mask and mane pommel and backpiece, with a carved ivory grip. As the dirk was unofficial, patterns were many and varied; but nearly all had in common the shortness of the blade and the ivory grip and gilt mounts.

Weapons of the ship's company

The ship's company were normally armed with cutlasses, pikes and boarding axes. The cutlass of this period had a straight blade some 26 inches long, terminating in a double spear point, and the guard consisted of a steel figure-of-eight shell, one circle of the eight being round the blade and the other in the middle of the guard. The grip was a tube of sheet steel roughly hammer-welded together and driven on to the tang of the blade, which itself was hammered over at the top of the guard. In about 1804, the grip appears to be of grooved cast iron driven on to the tang. The blade of the cutlass was normally marked with a crown and G.R.

The pike, which was mounted on a six-foot pole,

had a triangular head terminating in a very acute point, and the boarding axe, or tomahawk, had a double head with a curved-edged blade on one side and an acute spike on the other.

Two other weapons carried by a ship's company at that time were muskets and pistols. On some ships a seven-barrelled volley gun was carried. The muskets were made from parts of obsolete land service muskets. One point they had in common was the flat butt plate and in many cases the absence of a metal fore-end to the stock, as the muskets were not intended to take bayonets and so would not need to have the fore-end protected from the bayonet collar.

The pistol issued to ships' crews was termed the sea service pattern. This had a 12-inch barrel of 24 bore and was fitted with a standard lock. The stock was made of walnut and the butt was finished off with a plain rounded brass cap with short ears. On the reverse side to the lock was a long belt-hook and below the barrel and in the woodwork there were brass pipes which held the wooden ramrod.

In many cases these pistols can be found engraved with an identification mark on the brass butt cap; for example 'QD 24' standing for Quarter Deck pistol No 24.

In 1820 this type of pistol was superseded by a new pattern which was in fact a conversion of the long pistol. The barrel was reduced to nine inches and the ramrod was of steel, not wood.

The volley gun appears to have been suggested by a Captain James Wilson of the Royal Marines. This flintlock weapon had seven parallel barrels (six in a circle around the seventh) all arranged to fire at one time. The gun was adopted by the Navy for use by the men in the fighting tops, for clearing enemy decks and fighting tops.

Their manufacture was entrusted to one man only, Henry Nock, and he in fact made only two batches, 500 of the first model in 1780 and 100 of the second model, which differed in details of the lock, in 1787. Nothing is recorded about these volley guns in Ordnance records after 1790, and they were probably obsolescent by 1810.

GLOSSARY

Artificer. Skilled workman or craftsman.

Axle tree. The bar connecting opposite wheels of a carriage.

Base ring. First ring at the breech end of a barrel on which is usually marked the name of the caster.

Battering train. Ordnance and its equipment used for siege work.

Battery. A number of pieces of Ordnance and their equipment.

Block trail. The part of a gun carriage resting on the ground behind the wheels, made of one solid piece of wood.

Bomb ketch. A specially designed and armed ship for inshore bombardment of coastal targets.

Button. The round portion at the rear of a barrel.

Calibre. The diameter of the bore.

Canister shot. A tubular container loaded with a number of cast metal shot.

Cannon lock. A lock working on the same principle as a musket or pistol lock fitted to a piece of artillery to fire it.

Capsquares. Curved metal plates which lock over the trunnions of a barrel and keep it in place on a carriage.

Carcass. An incendiary shell.

Carronade. A short-barrelled piece made at the Carron works in Scotland and used particularly in the Navy and coastal fortifications.

Cascable. The part behind the base ring of a barrel.

Chamber. The rear end of the bore of a barrel.

Charge. The correct load of propellant for any particular piece of ordnance.

Chase. The portion of a barrel just before the muzzle.

Cheeks. Side pieces of a gun carriage on which the barrel rests.

Clinometer. Instrument for measuring the slope on which a piece of artillery stands when fitted to the piece.

Deadhead. Part of a cast barrel removed before finishing.

Dog wood. Wild cornel, a kind of flowering shrub.

Dolphins. The lifting handles on a brass barrel, so called because the early types were shaped like dolphins.

Fellies. Curved pieces of wood in the outer circumference of a wheel.

Friction tube. A tube primed with powder and used for igniting the main charge of a piece of artillery. The act of withdrawing a bar in the tube caused a spark by friction.

Fuse. Sometimes spelt fuze. A combustible cord for igniting a bomb or shell, or a component screwed or fitted into a shell for the same purpose.

Futchell. A lengthwise piece of timber in a carriage supporting the splinter bar.

Grape, grapeshot. So called because the balls, enclosed in a canvas bag, resembled a bunch of grapes.

Graze. The point where a projectile touches ground (or a building etc) without coming to rest.

Gyn. A tripod for raising a barrel from its carriage.

Handspike. A wooden lever used for moving a piece of artillery from side to side or for raising the barrel.

Howitzer. Short piece of ordnance for firing at high angles but at lower velocity than a gun.

Langet. Projection from a blade fitted to a pole by which the blade is secured. Also portion on the lower part of a sword hilt that projects over the top of the blade.

Limber. Detachable part of a gun carriage usually fitted with ammunition boxes.

Magazine. Place on a ship or on shore where powder and cartridges are stored.

Match. A piece of wick or cord designed to burn at a uniform rate.

Match tub. Tub, half filled with water, having notches in the rim in which matches were placed.

Mortar. A short piece of ordnance for throwing shells at very high angles.

Muzzle. The open end of the bore of a barrel, from which the projectile leaves the gun.

Nave. Central block of a wheel holding the axle and spokes.

Ordnance. A term used to describe all kinds of artillery.

Perch. Central pole of various carriages.

Piece. Term used to describe any gun, howitzer, mortar or carronade.

Platform. A construction of wood built under a gun to prevent the wheels from being driven into the ground by continuous firing.

Portfire. A fuse for igniting artillery.

Pound shot. A shot of 1 lb weight.

Priming. To place a small amount of powder on the vent of a piece which, when ignited, fired the main charge.

Prop stick. A stick used to prop up the shaft of a limber when standing without horses.

Quadrant. Quarter-circle instrument, of metal or wood, graduated to take angular measurements.

Quillon. The arm on the cross-guard of a sword hilt on the opposite side to the guard.

Quill tube. A tube of goose quill filled with powder and used in place of loose powder for priming a piece.

Quoin. A wooden wedge used for raising and depressing a barrel on its carriage.

Ramming. The act of forcing the projectile and charge down into the barrel.

Saltpetre. Potassium nitrate, or nitre.

Shabraque. Cavalry term for a saddle cloth.

Shrapnel. Name used to describe spherical case shot, taken from the name of the inventor, Lieutenant Henry Shrapnel.

Streak. Section of the metal tyre of a gun-carriage wheel.

Swivel gun. Light gun mounted by a bar into a hole, and capable of being swivelled in all directions. Mainly used at sea.

Tampion. A wooden stopper for the muzzle of a gun.

Tangent scale. A scale for measuring angles of elevation, fitted to the base ring of a barrel.

Traces. The straps by which horses are harnessed to a carriage.

Trail. The part of a gun carriage behind the wheels resting on the ground.

Train (noun). A collective name to describe the various guns, waggons and other equipment formed as one force of artillery.

(verb) To train or move a piece to bear on a target.

Trucks. Small wheels without spokes, used on gun carriages, made from solid wood or iron.

Trunnions. Horizontal round projections from the side of a barrel by which the barrel is mounted on a carriage and pivots upwards and downwards.

Vent. The small hole at the rear of the barrel through which the main charge is ignited.

Wad hook. A hook for removing wads from the barrel of a piece of artillery.

BIBLIOGRAPHY

Manuscript sources

Admiralty
Various orders and plans relative to naval ordnance
Author's collection
'Artillery Memorandums relative to the Royal Navy'
by Captain Robert Lawson, 1782
'Details of Experiments 1812, 1818, 1821',
anonymous
'Table raisonnee d'artillerie', by M. de Gribeauval,
1768, with additions 1774
Major J. Wilkinson-Latham
'Artillery notebook'; handwritten notebook of the
course at the Royal Military Academy, 1818
'Note book', by Henry Wilkinson, 1825
National Maritime Museum
Various plans and details of fireships and bomb
vessels
Public Record Office
Various minutes of the Board of Ordnance
Royal Artillery Institution
Various plans of guns and carriages and details of
performances of ordnance

Published sources

Blackmore, H. L. *British Military Firearms 1650–
    1850* (1961)
Boutell, Charles. *Arms and Armour* (1874)
Carman, W. Y. *A History of Firearms* (1955)
Congreve, William. *An Elementray Treatise on the
    mounting of Naval Ordnance* (1811)
——*The Details of the Rocket System* (1814)
Douglas, Major-General Sir Howard. *A Treatise on
    Naval Gunnery* (1820)
ffoulkes, Charles J. *The Gunfounders of England*
    (Cambridge, 1937)
——*Inventory and Survey of the Armouries of the
    Tower of London*, two vols (1916)
Field, Colonel C. *Britain's Sea Soldiers* (1924)
Gray, John. *A Treatise of Gunnery* (1731)
Greener, W. *The Science of Gunnery* (1841)
Hime, Henry W. *The Origins of Artillery* (1915)
Hughes, Major-General B. P. *British Smoothbore
    Artillery* (1969)
MacDonald, Captain R. J. *The History of the Dress
    of the Royal Regiment of Artillery* (1899)
May, Commander W. E. *Dress of Naval Officers*
    (1966)
May, Commander W. E., and Kennard, A. N. *Naval
    Swords and Firearms* (1962)
Napoleon Bonaparte, Charles Louis. *On Artillery*
    (Paris, 1851; reprinted London, 1967)
Pope, Dudley. *The Gun* (1965)
Robins, Benjamin. *New Principles of Gunnery*
    (1805 edition)
Rudyard, William. *Course of Artillery at the Royal
    Military Academy 1793* (Ottawa, 1969)
Wilkinson, Henry. *Engines of War* (1841)
Wilkinson-Latham, J. *British Cut and Thrust
    Weapons* (1971)
Wilson, Lieutenant A. W. *The Story of the Gun*
    (1944)

*Abridgement of the Patent Specifications Relating to
    Firearms and Other Weapons, Ammunition and
    Accoutrements from 1588 to 1858* (1858, reprint
    edition 1960)
*Catalogue of the Museum of Artillery in the Rotunda
    at Woolwich* (1963 edition)

ACKNOWLEDGEMENTS

My thanks are due to my father, Major J. Wilkinson-Latham, not only for his advice and for contributing the appendices on side arms but for the loan of the handwritten artillery notebook by a cadet at the Royal Military Academy, Woolwich, in 1818. I would also like to express my gratitude to the following, who all contributed much of their time in searching out for me either information, photographs or drawings from the collections in their charge: Major R. G. Bartelot and his staff of the Royal Artillery Institution, whose help was invaluable; P. G. W. Annis of the National Maritime Museum, who procured suitable photographs and read and corrected the text on naval uniforms; G. P. B. Naish, until recently the Keeper of the National Maritime Museum, who from his great knowledge explained the detailed workings of a ship of the line preparing for action; A. N. Kennard, until recently Assistant Master of the Tower of London Armouries for his help and advice and for selecting suitable illustrations; Boris Mollo and his staff of the National Army Museum for help with illustrations.

My final thanks must go to Jack Cassin Scott, who undertook research work for me besides doing many of the line drawings for the text, and to Jack Blake who took many of the photographs.

R. J. Wilkinson-Latham
Cannes
FRANCE

Picture Acknowledgements

Admiralty, 9, 10, 11
Aide Memoire to the Military Sciences (1845), 39, 40
Author's Collection, 2, 8, 19, 20, 21, 49, 59
From the Artillery Notebook of William Basset (Major J. Wilkinson-Latham Collection), 1, 3, 4, 25, 26, 27, 29, 35, 41, 42, 43, 45, 51, 52, 53, 54, 55, 57
Maggs Bros, London, 7
National Army Museum, London, 34, 36, 63
National Maritime Museum, Greenwich, 12, 60, 61
Parker Gallery, London, 6
Royal Army Museum, Stockholm, 33
Royal Artillery Institution, Woolwich, 5, 15, 17, 18, 28, 30, 32, 37, 47, 48, 58

Royal Marine Historical Photographic Library, Chatham, 62
Drawn by Jack Cassin Scott, 46, 50
Tower of London Armouries, 24
Drawn by Christine Wilkinson-Latham, 14, 31
Drawn by Major J. Wilkinson-Latham, 13, 16, 22, 23
Commanding Officer, HMS *Victory*, 44
Plates 5, 15, 17, 18, 28, 47, 48, 58 were photographed by Jack Blake at the Rotunda Museum, Woolwich by kind permission of the Royal Artillery Institution. Plates 1, 3, 4, 25, 26, 27, 29, 35, 41, 42, 43, 45, 57 were redrawn from the Artillery Notebook of William Basset by Jack Cassin Scott.

INDEX

Page numbers in italic type indicate illustrations

hand grenades, 33
Hawkins, Richard Francis, 83
harness, 15; *65*
Haycroft, Joseph, 83
Henry VIII, 7
Honourable East India Company, 21, 35
howitzers, 11–12, 18, 48, 85; *11, 50, 52, 56*

invalid companies, 7

James, Henry, 84
Jones, John, 84

Kildare, John, Marquis of, 7
King, John and Henry, 8
Kinman, Francis, 9
Koehler, Lieutenant G. F., 12

Langridge shot, 30, 32
laying of artillery, 43
light artillery, 9
limber, 54; *51, 54, 57*
light ball, 33–4
linstock, 38; *39*
Logan, Michael, 83

magazine at sea, 72
Maritz (gunfounder), 45
Melville, General Robert, 10
Moorfields foundry, *see* Bagley, Mathew
mortars, 12, 16, 77, 86, 87; *13*
——, beds for, 59, 75; *47, 78*
——, boring of, 48
——, drill with, 69
mould for casting ordnance, 44, 45
Muller, John, 42

Nock, Henry, 103–4

Parr, William, 83
patents for artillery 83–4
percussion fuse, 42
Poggi, Anthony Cesare de, 83
portfire, 38
projectile, 24–37; *see also under* specific headings
proof of ordnance, 49–51

quadrant, 43
quick match, 38

quill tubes, 38, 39
quoin, 9, 13, 73, 74

red-hot shot, 25–26
Richards, George, 84
Richmond, Duke of, 7
rockets, 34–7; *34, 35, 36*
Rocket Troop, 8, 14, 35; *95*
rope rammer, 43
round shot, 24–7; *25*
Royal Brass Foundry, 7, 8
Royal Horse Artillery, 7, 13, 53; *15, 95*
Royal Irish Artillery, 7
Royal Laboratory, Woolwich, 22
Royal Marine Artillery, 76; *102*

sabot, 23; *25*
Sadler's 'War Chariot', 14; *15*
saltpetre, 21
Schalch, Andrew, 8, 9
shells, oblong, experiments with, 18
Shrapnel, Lieutenant, Henry, 28, 29
sling cast, 70; *69*
slow match, 38
Smith, Henry, 83
smoke ball, 33–4
spherical case shot (shrapnel), 28; *25*

Tower of London, 7, 24, 41–2
traversing slide carriage, 18
trucks, 59; *18*
trunnions, cutting of, 48, 49
tubes, friction, 39
——, paper, 38, 39
——, quill, 38, 39
——, tin, 38

Uniforms, 92–107

Vazie, Robert, 84
vents, bushing of, 48; *48*
Verbruggen, Jan and Peter, 8

Walker, Humphrey, 7
Walker, James, 84
Western, Maximilian, 8
wheels, 54
——, construction of, 58, 59; *58*
Windsor, Frederick Albert, 84

Zimmerman, Samuel, 28, 29